Control Theory Fundamentals

Control Theory
Fundamentals

Edited by
Camden Howard

Larsen & Keller
www.larsen-keller.com

Control Theory Fundamentals
Edited by Camden Howard
ISBN: 978-1-63549-077-0 (Hardback)

Larsen & Keller

Published by Larsen and Keller Education,
5 Penn Plaza,
19th Floor,
New York, NY 10001, USA

Cataloging-in-Publication Data

Control theory fundamentals / edited by Camden Howard.
     p. cm.
Includes bibliographical references and index.
ISBN 978-1-63549-077-0
1. Automatic control. 2. Control theory. 3. Electronic control. 4. Electric controllers. I. Howard, Camden.
TJ213 .C66 2017
629.8--dc23

The publisher's policy is to use permanent paper from mills that operate a sustainable forestry policy. Furthermore, the publisher ensures that the text paper and cover boards used have met acceptable environmental accreditation standards.

Printed and bound in the United States of America.

For more information regarding Larsen and Keller Education and its products, please visit the publisher's website www.larsen-keller.com

Table of Contents

Permissions

Index

Preface

This book explores all the important aspects of control theory in the present day scenario. It provides thorough insights into this field and explains in detail the uses and methods of the subject. Control theory deals particularly with control systems or any other systems that produce a desired output through the input of references or values. It studies the behavior and inputs of dynamical systems. The book studies, analyses and uphold the pillars of this field and its utmost significance in modern times. It picks up individual branches and explains their need and contribution in the context of the growth of this subject and technology. The topics covered in this extensive book deal with the core subjects of this area. For all those who are interested in control theory, this textbook can prove to be an essential guide.

Given below is the chapter wise description of the book:

Chapter 1- Control theory combines the principles of engineering and mathematics in order to deal with the behavior of changeable systems and how the change is brought about. It is profoundly connected to control systems. This chapter explains to the reader the significance of control theory.

Chapter 2- This chapter gives an overview on control systems. A control system is a device that manages, commands or regulates the behavior of other systems or devices. Control systems can be applied to manual operations or machines that require or can facilitate an operator. The content on control systems offers an insightful focus, keeping in mind the complex subject matter.

Chapter 3- Control systems can best be understood in confluence with the major topics listed in the following chapter. The major categories of control systems are dealt with great detail in the chapter. Industrial control system, PID controller, fly by wire are some of the control systems analyzed in this chapter. The topics discussed are of great importance to broaden the existing knowledge on control systems.

Chapter 4- Tools and methods are an important component of any field of study. The following chapter elucidates the various tools and methods of control systems. Some of the tools, like a transducer, are explained in this chapter. The following text also explains to the reader the relevance of control systems.

Chapter 5- The application of control theory to design systems with desired behavior is certified as control engineering. This chapter is a compilation of the important topics related to control engineering, such as senor and acutuator. It offers an insightful focus, keeping in mind the complex subject matter.

Chapter 6- System architecture deals with the design of systems that store content or data. It may use different computer languages or architecture description languages to access its content. This chapter gives an in-depth understanding of system architecture, and provides the reader with an elucidated knowledge on the subject matter.

Chapter 7- The following chapter elucidates the applications that are related to control system. It discusses the functions of control systems in a critical manner providing key analysis to the subject matter. The applications explained are electrical network, digital signal processor, microcontroller and cruise control.

At the end, I would like to thank all those who dedicated their time and efforts for the successful completion of this book. I also wish to convey my gratitude towards my friends and family who supported me at every step.

Editor

Introduction to Control Theory

Control theory combines the principles of engineering and mathematics in order to deal with the behavior of changeable systems and how the change is brought about. It is profoundly connected to control systems. This chapter explains to the reader the significance of control theory.

Control Theory

Control theory is an interdisciplinary branch of engineering and mathematics that deals with the behavior of dynamical systems with inputs, and how their behavior is modified by feedback. The usual objective of control theory is to control a system, often called the *plant*, so its output follows a desired control signal, called the *reference*, which may be a fixed or changing value. To do this a *controller* is designed, which monitors the output and compares it with the reference. The difference between actual and desired output, called the *error* signal, is applied as feedback to the input of the system, to bring the actual output closer to the reference. Some topics studied in control theory are stability (whether the output will converge to the reference value or oscillate about it), controllability and observability.

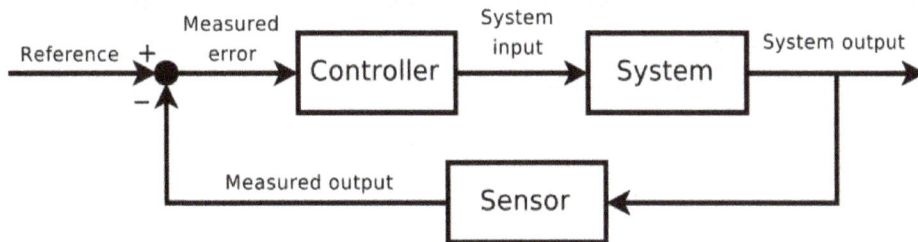

A block diagram of a negative feedback control system. Illustrates the concept of using a feedback loop to control the behavior of a system by comparing its output to a desired value, and applying the difference as an error signal to dynamically change the output so it is closer to the desired output

Extensive use is usually made of a diagrammatic style known as the block diagram. The transfer function, also known as the system function or network function, is a mathematical representation of the relation between the input and output based on the differential equations describing the system.

Although a major application of control theory is in control systems engineering, which deals with the design of process control systems for industry, other applications range far beyond this. As the general theory of feedback systems, control theory is useful wherever feedback occurs. A few examples are in physiology, electronics, climate modeling, machine design, ecosystems, navigation, neural networks, predator–prey interaction, gene expression, and production theory.

Overview

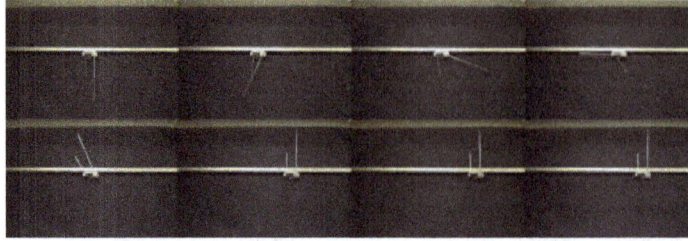

Smooth nonlinear trajectory planning with linear quadratic Gaussian feedback (LQR)
control on a dual pendula system.

Control theory is

- a theory that deals with influencing the behavior of dynamical systems

- an interdisciplinary subfield of science, which originated in engineering and mathematics, and evolved into use by the social sciences, such as economics, psychology, sociology, criminology and in the financial system.

Control systems may be thought of as having four functions: measure, compare, compute and correct. These four functions are completed by five elements: detector, transducer, transmitter, controller and final control element. The measuring function is completed by the detector, transducer and transmitter. In practical applications these three elements are typically contained in one unit. A standard example of a measuring unit is a resistance thermometer. The compare and compute functions are completed within the controller, which may be implemented electronically by proportional control, a PI controller, PID controller, bistable, hysteretic control or programmable logic controller. Older controller units have been mechanical, as in a centrifugal governor or a carburetor. The correct function is completed with a final control element. The final control element changes an input or output in the control system that affects the manipulated or controlled variable.

An Example

An example of a control system is a car's cruise control, which is a device designed to maintain vehicle speed at a constant *desired* or *reference* speed provided by the driver. The *controller* is the cruise control, the *plant* is the car, and the *system* is the car and the cruise control. The system output is the car's speed, and the control itself is the engine's throttle position which determines how much power the engine delivers.

A primitive way to implement cruise control is simply to lock the throttle position when the driver engages cruise control. However, if the cruise control is engaged on a stretch of flat road, then the car will travel slower going uphill and faster when going downhill. This type of controller is called an *open-loop controller* because there is no feedback; no measurement of the system output (the car's speed) is used to alter the control (the throttle position.) As a result, the controller cannot compensate for changes acting on the car, like a change in the slope of the road.

In a *closed-loop control system*, data from a sensor monitoring the car's speed (the system output) enters a controller which continuously subtracts the quantity representing the speed from the reference quantity representing the desired speed. The difference, called the error, determines the throttle position (the control). The result is to match the car's speed to the reference speed (maintain the desired system output). Now, when the car goes uphill, the difference between the input (the sensed speed) and the reference continuously determines the throttle position. As the sensed speed drops below the reference, the difference increases, the throttle opens, and engine power increases, speeding up the vehicle. In this way, the controller dynamically counteracts changes to the car's speed. The central idea of these control systems is the *feedback loop*, the controller affects the system output, which in turn is measured and fed back to the controller.

Classification

Linear Versus Nonlinear Control Theory

The field of control theory can be divided into two branches:

- *Linear control theory* – This applies to systems made of devices which obey the superposition principle, which means roughly that the output is proportional to the input. They are governed by linear differential equations. A major subclass is systems which in addition have parameters which do not change with time, called *linear time invariant* (LTI) systems. These systems are amenable to powerful frequency domain mathematical techniques of great generality, such as the Laplace transform, Fourier transform, Z transform, Bode plot, root locus, and Nyquist stability criterion. These lead to a description of the system using terms like bandwidth, frequency response, eigenvalues, gain, resonant frequencies, poles, and zeros, which give solutions for system response and design techniques for most systems of interest.

- *Nonlinear control theory* – This covers a wider class of systems that do not obey the superposition principle, and applies to more real-world systems, because all real control systems are nonlinear. These systems are often governed by nonlinear differential equations. The few mathematical techniques which have been developed to handle them are more difficult and much less general, often applying only to narrow categories of systems. These include limit cycle theory, Poincaré maps, Lyapunov stability theorem, and describing functions. Nonlinear systems are often analyzed using numerical methods on computers, for example by simulating their operation using a simulation language. If only solutions near a stable point are of interest, nonlinear systems can often be linearized by approximating them by a linear system using perturbation theory, and linear techniques can be used.

Frequency Domain Versus Time Domain

Mathematical techniques for analyzing and designing control systems fall into two different categories:

- *Frequency domain* – In this type the values of the state variables, the mathematical variables representing the system's input, output and feedback are represented as functions of frequency. The input signal and the system's transfer function are converted from time functions to functions of frequency by a transform such as the Fourier transform, Laplace transform, or Z transform. The advantage of this technique is that it results in a simplification of the mathematics; the *differential equations* that represent the system are replaced by *algebraic equations* in the frequency domain which are much simpler to solve. However, frequency domain techniques can only be used with linear systems, as mentioned above.

- *Time-domain state space representation* – In this type the values of the state variables are represented as functions of time. With this model the system being analyzed is represented by one or more differential equations. Since frequency domain techniques are limited to linear systems, time domain is widely used to analyze real-world nonlinear systems. Although these are more difficult to solve, modern computer simulation techniques such as simulation languages have made their analysis routine.

Siso Vs Mimo

Control systems can be divided into different categories depending on the number of inputs and outputs.

- Single-input single-output (SISO) – This is the simplest and most common type, in which one output is controlled by one control signal. Examples are the cruise control example above, or an audio system, in which the control input is the input audio signal and the output is the sound waves from the speaker.

- Multiple-input multiple-output (MIMO) – These are found in more complicated systems. For example, modern large telescopes such as the Keck and MMT have mirrors composed of many separate segments each controlled by an actuator. The shape of the entire mirror is constantly adjusted by a MIMO active optics control system using input from multiple sensors at the focal plane, to compensate for changes in the mirror shape due to thermal expansion, contraction, stresses as it is rotated and distortion of the wavefront due to turbulence in the atmosphere. Complicated systems such as nuclear reactors and human cells are simulated by computer as large MIMO control systems.

History

Centrifugal governor in a Boulton & Watt engine of 1788

Although control systems of various types date back to antiquity, a more formal analysis of the field began with a dynamics analysis of the centrifugal governor, conducted by the physicist James Clerk Maxwell in 1868, entitled *On Governors*. This described and analyzed the phenomenon of self-oscillation, in which lags in the system may lead to overcompensation and unstable behavior. This generated a flurry of interest in the topic, during which Maxwell's classmate, Edward John Routh, abstracted Maxwell's results for the general class of linear systems. Independently, Adolf Hurwitz analyzed system stability using differential equations in 1877, resulting in what is now known as the Routh–Hurwitz theorem.

A notable application of dynamic control was in the area of manned flight. The Wright brothers made their first successful test flights on December 17, 1903 and were distinguished by their ability to control their flights for substantial periods (more so than the ability to produce lift from an airfoil, which was known). Continuous, reliable control of the airplane was necessary for flights lasting longer than a few seconds.

By World War II, control theory was an important part of fire-control systems, guidance systems and electronics.

Sometimes, mechanical methods are used to improve the stability of systems. For example, ship stabilizers are fins mounted beneath the waterline and emerging laterally. In contemporary vessels, they may be gyroscopically controlled active fins, which have the capacity to change their angle of attack to counteract roll caused by wind or waves acting on the ship.

The Sidewinder missile uses small control surfaces placed at the rear of the missile with spinning disks on their outer surfaces and these are known as rollerons. Airflow over the disks spins them to a high speed. If the missile starts to roll, the gyroscopic force of the disks drives the control surface into the airflow, cancelling the motion. Thus, the Sidewinder team replaced a potentially complex control system with a simple mechanical solution.

The Space Race also depended on accurate spacecraft control, and control theory has also seen an increasing use in fields such as economics.

People in Systems and Control

Many active and historical figures made significant contribution to control theory including

- Pierre-Simon Laplace (1749–1827) invented the Z-transform in his work on probability theory, now used to solve discrete-time control theory problems. The Z-transform is a discrete-time equivalent of the Laplace transform which is named after him.

- Alexander Lyapunov (1857–1918) in the 1890s marks the beginning of stability theory.

- Harold S. Black (1898–1983), invented the concept of negative feedback amplifiers in 1927. He managed to develop stable negative feedback amplifiers in the 1930s.

- Harry Nyquist (1889–1976) developed the Nyquist stability criterion for feedback systems in the 1930s.

- Richard Bellman (1920–1984) developed dynamic programming since the 1940s.

- Andrey Kolmogorov (1903–1987) co-developed the Wiener–Kolmogorov filter in 1941.

- Norbert Wiener (1894–1964) co-developed the Wiener–Kolmogorov filter and coined the term cybernetics in the 1940s.

- John R. Ragazzini (1912–1988) introduced digital control and the use of Z-transform in control theory (invented by Laplace) in the 1950s.

- Lev Pontryagin (1908–1988) introduced the maximum principle and the bang-bang principle.

- Pierre-Louis Lions (1956) developed viscosity solutions into stochastic control and optimal control methods.

Classical Control Theory

To overcome the limitations of the open-loop controller, control theory introduces feedback. A closed-loop controller uses feedback to control states or outputs of a dynamical system. Its name comes from the information path in the system: process inputs (e.g., voltage applied to an electric motor) have an effect on the process outputs (e.g., speed or torque of the motor), which is measured with sensors and processed by the controller; the result (the control signal) is "fed back" as input to the process, closing the loop.

Closed-loop controllers have the following advantages over open-loop controllers:

- disturbance rejection (such as hills in the cruise control example above)

- guaranteed performance even with model uncertainties, when the model structure does not match perfectly the real process and the model parameters are not exact

- unstable processes can be stabilized

- reduced sensitivity to parameter variations

- improved reference tracking performance

In some systems, closed-loop and open-loop control are used simultaneously. In such systems, the open-loop control is termed feedforward and serves to further improve reference tracking performance.

A common closed-loop controller architecture is the PID controller.

Closed-loop Transfer Function

The output of the system $y(t)$ is fed back through a sensor measurement F to a comparison with the reference value $r(t)$. The controller C then takes the error e (difference) between the reference and the output to change the inputs u to the system under control P. This is shown in the figure. This kind of controller is a closed-loop controller or feedback controller.

This is called a single-input-single-output (*SISO*) control system; *MIMO* (i.e., Multi-Input-Multi-Output) systems, with more than one input/output, are common. In such cases variables are represented through vectors instead of simple scalar values. For some distributed parameter systems the vectors may be infinite-dimensional (typically functions).

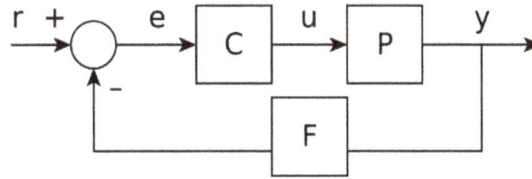

If we assume the controller C, the plant P, and the sensor F are linear and time-invariant (i.e., elements of their transfer function $C(s)$, $P(s)$, and $F(s)$ do not depend on time), the systems above can be analysed using the Laplace transform on the variables. This gives the following relations:

$$Y(s) = P(s)U(s)$$

$$U(s) = C(s)E(s)$$

$$E(s) = R(s) - F(s)Y(s).$$

Solving for $Y(s)$ in terms of $R(s)$ gives

$$Y(s) = \left(\frac{P(s)C(s)}{1 + F(s)P(s)C(s)} \right) R(s) = H(s)R(s).$$

The expression $H(s) = \dfrac{P(s)C(s)}{1 + F(s)P(s)C(s)}$

is referred to as the *closed-loop transfer function* of the system. The numerator is the forward (open-loop) gain from r to y, and the denominator is one plus the gain in going around the feedback loop, the so-called loop gain. If $|P(s)C(s)| \gg 1$, , i.e., it has a large norm with each value of s, and if $|F(s)| \approx 1$, , then $Y(s)$ is approximately equal to $R(s)$ and the output closely tracks the reference input.

PID Controller

The PID controller is probably the most-used feedback control design. *PID* is an initialism for *Proportional-Integral-Derivative*, referring to the three terms operating on the error signal to produce a control signal. If $u(t)$ is the control signal sent to the system, $y(t)$ is the measured output and $r(t)$ is the desired output, and tracking error $e(t) = r(t) - y(t)$, a PID controller has the general form

$$u(t) = K_P e(t) + K_I \int e(t)dt + K_D \frac{d}{dt} e(t).$$

The desired closed loop dynamics is obtained by adjusting the three parameters K_P , K_I and K_D , often iteratively by "tuning" and without specific knowledge of a plant model. Stability can often be ensured using only the proportional term. The integral term permits the rejection of a step disturbance (often a striking specification in process control). The derivative term is used to provide damping or shaping of the response. PID controllers are the most well established class of control systems: however, they cannot be used in several more complicated cases, especially if MIMO systems are considered.

Applying Laplace transformation results in the transformed PID controller equation

$$u(s) = K_P e(s) + K_I \frac{1}{s} e(s) + K_D s e(s)$$

$$u(s) = \left(K_P + K_I \frac{1}{s} + K_D s \right) e(s)$$

with the PID controller transfer function

$$C(s) = \left(K_P + K_I \frac{1}{s} + K_D s \right).$$

There exists a nice example of the closed-loop system discussed above. If we take

PID controller transfer function in series form

$$C(s) = K \left(1 + \frac{1}{sT_i} \right)(1 + sT_d)$$

1st order filter in feedback loop

$$F(s) = \frac{1}{1 + sT_f}$$

linear actuator with filtered input

$$P(s) = \frac{A}{1 + sT_p} \ , \ A = \text{const}$$

and insert all this into expression for closed-loop transfer function H(s), then tuning is very easy: simply put

$$K = \frac{1}{A}, T_i = T_f, T_d = T_p$$

and get H(s) = 1 identically.

For practical PID controllers, a pure differentiator is neither physically realisable nor desirable due to amplification of noise and resonant modes in the system. Therefore, a phase-lead compensator type approach is used instead, or a differentiator with low-pass roll-off.

Modern Control Theory

In contrast to the frequency domain analysis of the classical control theory, modern control theory utilizes the time-domain state space representation, a mathematical model of a physical system as a set of input, output and state variables related by first-order differential equations. To abstract from the number of inputs, outputs and states, the variables are expressed as vectors and the differential and algebraic equations are written in matrix form (the latter only being possible when the dynamical system is linear). The state space representation (also known as the "time-domain approach") provides a convenient and compact way to model and analyze systems with multiple inputs and outputs. With inputs and outputs, we would otherwise have to write down Laplace transforms to encode all the information about a system. Unlike the frequency domain approach, the use of the state-space representation is not limited to systems with linear components and zero initial conditions. "State space" refers to the space whose axes are the state variables. The state of the system can be represented as a point within that space.

Topics in Control Theory

Stability

The *stability* of a general dynamical system with no input can be described with Lyapunov stability criteria.

- A linear system is called bounded-input bounded-output (BIBO) stable if its output will stay bounded for any bounded input.

- Stability for nonlinear systems that take an input is input-to-state stability (ISS), which combines Lyapunov stability and a notion similar to BIBO stability.

For simplicity, the following descriptions focus on continuous-time and discrete-time linear systems.

Mathematically, this means that for a causal linear system to be stable all of the poles of its transfer function must have negative-real values, i.e. the real part of each pole must be less than zero. Practically speaking, stability requires that the transfer function complex poles reside

- in the open left half of the complex plane for continuous time, when the Laplace transform is used to obtain the transfer function.

- inside the unit circle for discrete time, when the Z-transform is used.

The difference between the two cases is simply due to the traditional method of plotting continuous time versus discrete time transfer functions. The continuous Laplace transform is in Cartesian coordinates where the x axis is the real axis and the discrete Z-transform is in circular coordinates where the ρ axis is the real axis.

When the appropriate conditions above are satisfied a system is said to be asymptotically stable; the variables of an asymptotically stable control system always decrease from their initial value and do not show permanent oscillations. Permanent oscillations occur when a pole has a real part exactly equal to zero (in the continuous time case) or a modulus equal to one (in the discrete time case). If a simply stable system response neither decays nor grows over time, and has no oscillations, it is marginally stable; in this case the system transfer function has non-repeated poles at complex plane origin (i.e. their real and complex component is zero in the continuous time case). Oscillations are present when poles with real part equal to zero have an imaginary part not equal to zero.

If a system in question has an impulse response of

$$x[n] = 0.5^n u[n]$$

then the Z-transform (see this example), is given by

$$X(z) = \frac{1}{1 - 0.5z^{-1}}$$

which has a pole in $z = 0.5$ (zero imaginary part). This system is BIBO (asymptotically) stable since the pole is *inside* the unit circle.

However, if the impulse response was

$$x[n] = 1.5^n u[n]$$

then the Z-transform is

$$X(z) = \frac{1}{1 - 1.5z^{-1}}$$

which has a pole at $z = 1.5$ and is not BIBO stable since the pole has a modulus strictly greater than one.

Numerous tools exist for the analysis of the poles of a system. These include graphical systems like the root locus, Bode plots or the Nyquist plots.

Mechanical changes can make equipment (and control systems) more stable. Sailors add ballast to improve the stability of ships. Cruise ships use antiroll fins that extend transversely from the side of the ship for perhaps 30 feet (10 m) and are continuously rotated about their axes to develop forces that oppose the roll.

Controllability and Observability

Controllability and observability are main issues in the analysis of a system before deciding the best control strategy to be applied, or whether it is even possible to control or stabilize the system. Controllability is related to the possibility of forcing the system into a particular state by using an appropriate control signal. If a state is not controllable, then no signal will ever be able to control the state. If a state is not controllable, but its dynamics are stable, then the state is termed *stabilizable*. Observability instead is related to the possibility of *observing*, through output measurements, the state of a system. If a state is not observable, the controller will never be able to determine the behaviour of an unobservable state and hence cannot use it to stabilize the system. However, similar to the stabilizability condition above, if a state cannot be observed it might still be detectable.

From a geometrical point of view, looking at the states of each variable of the system to be controlled, every "bad" state of these variables must be controllable and observable to ensure a good behaviour in the closed-loop system. That is, if one of the eigenvalues of the system is not both controllable and observable, this part of the dynamics will remain untouched in the closed-loop system. If such an eigenvalue is not stable, the dynamics of this eigenvalue will be present in the closed-loop system which therefore will be unstable. Unobservable poles are not present in the transfer function realization of a state-space representation, which is why sometimes the latter is preferred in dynamical systems analysis.

Solutions to problems of uncontrollable or unobservable system include adding actuators and sensors.

Control Specification

Several different control strategies have been devised in the past years. These vary from extremely general ones (PID controller), to others devoted to very particular classes of systems (especially robotics or aircraft cruise control).

A control problem can have several specifications. Stability, of course, is always present. The controller must ensure that the closed-loop system is stable, regardless of the open-loop stability. A poor choice of controller can even worsen the stability of the open-loop system, which must normally be avoided. Sometimes it would be desired to obtain particular dynamics in the closed loop: i.e. that the poles have $Re[\lambda] < -\bar{\lambda}$, where $\bar{\lambda}$ is a fixed value strictly greater than zero, instead of simply asking that $Re[\lambda] < 0$..

Another typical specification is the rejection of a step disturbance; including an integrator in the open-loop chain (i.e. directly before the system under control) easily achieves this. Other classes of disturbances need different types of sub-systems to be included.

Other "classical" control theory specifications regard the time-response of the closed-loop system. These include the rise time (the time needed by the control system to reach the desired value after a perturbation), peak overshoot (the highest value reached by the response before reaching the desired value) and others (settling time, quarter-decay). Frequency domain specifications are usually related to robustness.

Modern performance assessments use some variation of integrated tracking error (IAE,ISA,CQI).

Model Identification and Robustness

A control system must always have some robustness property. A robust controller is such that its properties do not change much if applied to a system slightly different from the mathematical one used for its synthesis. This specification is important, as no real physical system truly behaves like the series of differential equations used to represent it mathematically. Typically a simpler mathematical model is chosen in order to simplify calculations, otherwise the true system dynamics can be so complicated that a complete model is impossible.

System Identification

The process of determining the equations that govern the model's dynamics is called system identification. This can be done off-line: for example, executing a series of measures from which to calculate an approximated mathematical model, typically its transfer function or matrix. Such identification from the output, however, cannot take account of unobservable dynamics. Sometimes the model is built directly starting from known physical equations, for example, in the case of a mass-spring-damper system we know that $m\ddot{x}(t) = -Kx(t) - B\dot{x}(t)$. . Even assuming that a "complete" model is used in designing the controller, all the parameters included in these equations (called "nominal parameters") are never known with absolute precision; the control system will have to behave correctly even when connected to physical system with true parameter values away from nominal.

Some advanced control techniques include an "on-line" identification process. The parameters of the model are calculated ("identified") while the controller itself is running. In this way, if a drastic variation of the parameters ensues, for example, if the robot's arm releases a weight, the controller will adjust itself consequently in order to ensure the correct performance.

Analysis

Analysis of the robustness of a SISO (single input single output) control system can be performed in the frequency domain, considering the system's transfer function and using Nyquist and Bode diagrams. Topics include gain and phase margin and amplitude margin. For MIMO (multi input multi output) and, in general, more complicated control systems one must consider the theoretical results devised for each control technique. i.e., if particular robustness qualities are needed, the engineer must shift his attention to a control technique by including them in its properties.

Constraints

A particular robustness issue is the requirement for a control system to perform properly in the presence of input and state constraints. In the physical world every signal is limited. It could happen that a controller will send control signals that cannot be followed by the physical system, for example, trying to rotate a valve at excessive speed. This can produce undesired behavior of the closed-loop system, or even damage or break actuators or other subsystems. Speci ic control techniques are available to solve the problem: model predictive control, and antiwind up systems. The latter consists of an additional control block that ensures that the control signal never exceeds a given threshold.

System Classifications

Linear Systems Control

For MIMO systems, pole placement can be performed mathematically using a state space representation of the open-loop system and calculating a feedback matrix assigning poles in the desired positions. In complicated systems this can require computer-assisted calculation capabilities, and cannot always ensure robustness. Furthermore, all system states are not in general measured and so observers must be included and incorporated in pole placement design.

Nonlinear Systems Control

Processes in industries like robotics and the aerospace industry typically have strong nonlinear dynamics. In control theory it is sometimes possible to linearize such classes of systems and apply linear techniques, but in many cases it can be necessary to devise from scratch theories permitting control of nonlinear systems. These, e.g., feedback linearization, backstepping, sliding mode control, trajectory linearization control normally take advantage of results based on Lyapunov's theory. Differential geometry has been widely used as a tool for generalizing well-known linear control concepts to the non-linear case, as well as showing the subtleties that make it a more challenging problem. Control theory has also been used to decipher the neural mechanism that directs cognitive states.

Decentralized Systems Control

When the system is controlled by multiple controllers, the problem is one of decentralized control. Decentralization is helpful in many ways, for instance, it helps control systems to operate over a larger geographical area. The agents in decentralized control systems can interact using communication channels and coordinate their actions.

Deterministic and Stochastic Systems Control

A stochastic control problem is one in which the evolution of the state variables is subjected to random shocks from outside the system. A deterministic control problem is not subject to external random shocks.

Main Control Strategies

Every control system must guarantee first the stability of the closed-loop behavior. For linear systems, this can be obtained by directly placing the poles. Non-linear control systems use specific theories (normally based on Aleksandr Lyapunov's Theory) to ensure stability without regard to the inner dynamics of the system. The possibility to fulfill different specifications varies from the model considered and the control strategy chosen.

List of the main control techniques

- Adaptive control uses on-line identification of the process parameters, or modification of controller gains, thereby obtaining strong robustness properties. Adaptive controls were applied for the first time in the aerospace industry in the 1950s, and have found particular success in that field.

- A hierarchical control system is a type of control system in which a set of devices and governing software is arranged in a hierarchical tree. When the links in the tree are implemented by a computer network, then that hierarchical control system is also a form of networked control system.

- Intelligent control uses various AI computing approaches like neural networks, Bayesian probability, fuzzy logic, machine learning, evolutionary computation and genetic algorithms to control a dynamic system.

- Optimal control is a particular control technique in which the control signal optimizes a certain "cost index": for example, in the case of a satellite, the jet thrusts needed to bring it to desired trajectory that consume the least amount of fuel. Two optimal control design methods have been widely used in industrial applications, as it has been shown they can guarantee closed-loop stability. These are Model Predictive Control (MPC) and linear-quadratic-Gaussian control (LQG). The first can more explicitly take into account constraints on the signals in the system, which is an important feature in many industrial processes. However, the "optimal control" structure in MPC is only a means to achieve such a result, as it does not optimize a true performance index of the closed-loop control system. Together with PID controllers, MPC systems are the most widely used control technique in process control.

- Robust control deals explicitly with uncertainty in its approach to controller design. Controllers designed using *robust control* methods tend to be able to cope with small differences between the true system and the nominal model used for design. The early methods of Bode and others were fairly robust; the state-space methods invented in the 1960s and 1970s were sometimes found to lack robustness. Examples of modern robust control techniques include H-infinity loop-shaping developed by Duncan McFarlane and Keith Glover of Cambridge University, United Kingdom and Sliding mode control (SMC) developed by Vadim Utkin. Robust methods aim to achieve robust performance and/or stability in the presence of small modeling errors.

- Stochastic control deals with control design with uncertainty in the model. In typical stochastic control problems, it is assumed that there exist random noise and disturbances in the model and the controller, and the control design must take into account these random deviations.

- Energy-shaping control view the plant and the controller as energy-transformation devices. The control strategy is formulated in terms of interconnection (in a power-preserving manner) in order to achieve a desired behavior.

- Self-organized criticality control may be defined as attempts to interfere in the processes by which the self-organized system dissipates energy.

Area & Approaches of Control System

Nonlinear Control

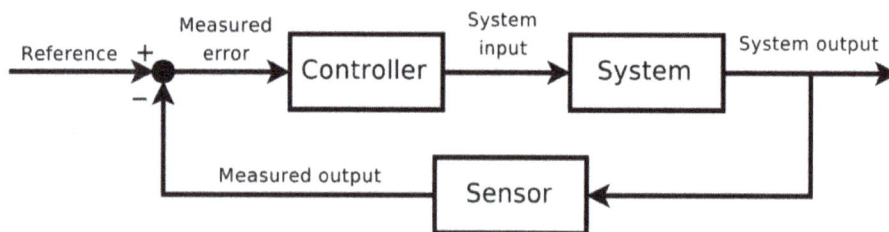

A feedback control system. It is desired to control a system (often called the *plant*) so its output follows a desired *reference* signal. A *sensor* monitors the output and a *controller* subtracts the actual output from the desired reference output, and applies this error signal to the system to bring the output closer to the reference. In a nonlinear control system at least one of the blocks, system, sensor, or controller, is nonlinear.

Nonlinear control theory is the area of control theory which deals with systems that are nonlinear, time-variant, or both. *Control theory* is an interdisciplinary branch of engineering and mathematics that is concerned with the behavior of dynamical systems with inputs, and how to modify the output by changes in the input using feedback. The system to be controlled is called the "plant". In order to make the output of a system follow a desired reference signal a controller is designed which compares the output of the plant to the desired output, and provides feedback to the plant to modify the output to bring it closer to the desired output. Control theory is divided into two branches:

Linear control theory applies to systems made of linear devices; which means they obey the superposition principle; the output of the device is proportional to its input. Systems with this property are governed by linear differential equations. A major subclass is systems which in addition have parameters which do not change with time, called *linear time invariant* (LTI) systems. These systems are amenable to powerful frequency domain mathematical techniques of great generality, such as the Laplace transform, Fourier transform, and Z transform, root-locus, Bode plot, and Nyquist stability criterion. These lead to a description of the system using terms like bandwidth, frequency response, eigenvalues, gain, resonant frequencies, poles, and zeros, which give solutions for system response and design techniques to most problems of interest.

Nonlinear control theory covers a wider class of systems that do not obey the superposition principle. It applies to more real-world systems, because all real control systems are nonlinear. These systems are often governed by nonlinear differential equations. The mathematical techniques which have been developed to handle them are more rigorous and much less general, often applying only to narrow categories of systems. These include limit cycle theory, Poincaré maps, Liapunov stability theory, and describing functions. If only solutions near a stable point are of interest, nonlinear systems can often be linearized by approximating them by a linear system obtained by expanding the nonlinear solution in a series, and then linear techniques can be used. Nonlinear systems are often analyzed using numerical methods on computers, for example by simulating their operation using a simulation language. Even if the plant is linear, a nonlinear controller can often have attractive features such as simpler implementation, faster speed, more accuracy, or reduced control energy, which justify the more difficult design procedure.

An example of a nonlinear control system is a thermostat-controlled heating system. A building heating system such as a furnace has a nonlinear response to changes in temperature; it is either "on" or "off", it does not have the fine control in response to temperature differences that a proportional (linear) device would have. Therefore, the furnace is off until the temperature falls below the "turn on" setpoint of the thermostat, when it turns on. Due to the heat added by the furnace, the temperature increases until it reaches the "turn off" setpoint of the thermostat, which turns the furnace off, and the cycle repeats. This cycling of the temperature about the desired temperature is called a *limit cycle*, and is characteristic of nonlinear control systems.

Properties of Conlinear Systems

Some properties of nonlinear dynamic systems are

- They do not follow the principle of superposition (linearity and homogeneity).

- They may have multiple isolated equilibrium points.

- They may exhibit properties such as limit cycle, bifurcation, chaos.

- Finite escape time: Solutions of nonlinear systems may not exist for all times.

Analysis and Control of Nonlinear Systems

There are several well-developed techniques for analyzing nonlinear feedback systems:

- Describing function method

- Phase plane method

- Lyapunov stability analysis

- Singular perturbation method

- Popov criterion (described in *The Lur'e Problem* below)

- Center manifold theorem

- Small-gain theorem

- Passivity analysis

Control design techniques for nonlinear systems also exist. These can be subdivided into techniques which attempt to treat the system as a linear system in a limited range of operation and use (well-known) linear design techniques for each region:

- Gain scheduling

Those that attempt to introduce auxiliary nonlinear feedback in such a way that the system can be treated as linear for purposes of control design:

- Feedback linearization

And Lyapunov based methods:

- Lyapunov redesign

- Nonlinear damping

- Backstepping

- Sliding mode control

Nonlinear Feedback Analysis – the Lur'e Problem

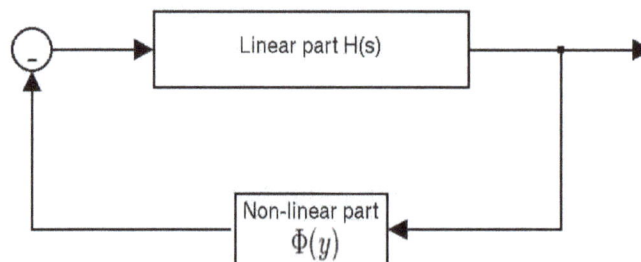

Lur'e problem block diagram

An early nonlinear feedback system analysis problem was formulated by A. I. Lur'e. Control systems described by the Lur'e problem have a forward path that is linear and time-invariant, and a feedback path that contains a memory-less, possibly time-varying, static nonlinearity.

The linear part can be characterized by four matrices (A,B,C,D), while the nonlinear part is $\Phi(y)$ with $\dfrac{\Phi(y)}{y} \in [a,b], \quad a < b \quad \forall y$ (a sector nonlinearity).

Absolute Stability Problem

Consider:

1. (A,B) is controllable and (C,A) is observable

2. two real numbers a, b with a<b, defining a sector for function Φ

The problem is to derive conditions involving only the transfer matrix H(s) and {a,b} such that x=0 is a globally uniformly asymptotically stable equilibrium of the system. This is known as the Lur'e problem. There are two well-known wrong conjections on absolute stability:

- The Aizerman's conjecture

- The Kalman's conjecture.

There are counterexamples to Aizerman's and Kalman's conjectures such that nonlinearity belongs to the sector of linear stability and unique stable equilibrium coexists with a stable periodic solution -- hidden oscillation.

There are two main theorems concerning the problem:

- The Circle criterion

- The Popov criterion.

which give sufficient conditions of absolute stability.

Popov Criterion

The sub-class of Lur'e systems studied by Popov is described by:

$$\begin{aligned}
\dot{x} &= Ax + bu \\
\dot{\xi} &= u \\
y &= cx + d\xi \quad (1)
\end{aligned}$$

$$u = -\phi(y) \quad (2)$$

where $x \in R^n$, ξ, u, y are scalars and A,b,c,d have commensurate dimensions. The nonlinear element Φ: R → R is a time-invariant nonlinearity belonging to *open sector* (0, ∞). This means that

$$\Phi(0) = 0, \; y\,\Phi(y) > 0, \; \forall\, y \neq 0;$$

The transfer function from u to y is given by

$$H(s) = \frac{d}{s} + c(sI - A)^{-1}b$$

Theorem: Consider the system (1)-(2) and suppose

1. A is Hurwitz

2. (A,b) is controllable

3. (A,c) is observable

4. d>0 and

5. $\Phi \in (0,\infty)$

then the system is globally asymptotically stable if there exists a number r>0 such that $\inf_{\omega \in R} \text{Re}[(1+j\omega r)h(j\omega)] > 0$.

Things to be noted:

- The Popov criterion is applicable only to autonomous systems

- The system studied by Popov has a pole at the origin and there is no direct pass-through from input to output

- The nonlinearity Φ must satisfy an open sector condition

Theoretical Results in Nonlinear Control

Frobenius Theorem

The Frobenius theorem is a deep result in Differential Geometry. When applied to Nonlinear Control, it says the following: Given a system of the form

$$\dot{x} = \sum_{i=1}^{k} f_i(x)u_i(t)$$

where $x \in R^n$, $f_1,\ldots,f_k$ are vector fields belonging to a distribution   and $u_i(t)$ are control functions, the integral curves of x are restricted to a manifold of dimension m if $\text{span}(\Delta) = m$ and Δ is an involutive distribution.

Feedback Linearization

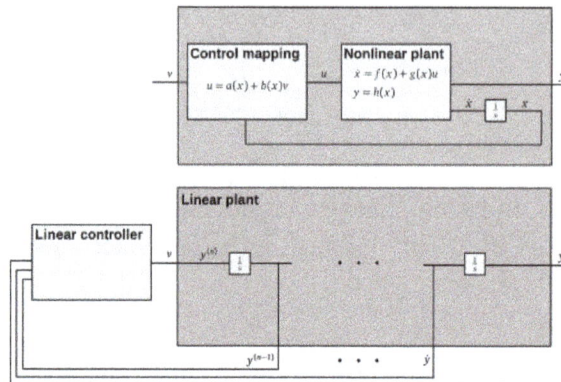

Block diagram illustrating the feedback linearization of a nonlinear system

Feedback linearization is a common approach used in controlling nonlinear systems. The approach involves coming up with a transformation of the nonlinear system into an equivalent linear system through a change of variables and a suitable control input. Feedback linearization may be applied to nonlinear systems of the form

$$\dot{x} = f(x)+g(x)u \qquad (1)$$
$$y = h(x) \qquad (2)$$

where $x \in \mathbb{R}^n$ is the state vector, $u \in \mathbb{R}^p$ is the vector of inputs, and $y \in \mathbb{R}^m$ is the vector of outputs. The goal is to develop a control input

$$u = a(x) + b(x)v$$

that renders a linear input–output map between the new input v and the output. An outer-loop control strategy for the resulting linear control system can then be applied.

Feedback Linearization of SISO Systems

Here, we consider the case of feedback linearization of a single-input single-output (SISO) system. Similar results can be extended to multiple-input multiple-output (MIMO) systems. In this case, $u \in \mathbb{R}$ and $y \in \mathbb{R}$. We wish to find a coordinate transformation $z = T(x)$ that transforms our system (1) into the so-called normal form which will reveal a feedback law of the form

$$u = a(x) + b(x)v$$

that will render a linear input–output map from the new input $v \in \mathbb{R}$ to the output y. To ensure that the transformed system is an equivalent representation of the original system, the transformation must be a diffeomorphism. That is, the transformation must not only be invertible (i.e., bijective), but both the transformation and its inverse must be smooth so that differentiability in the original coordinate system is preserved in the new coordinate system. In practice, the transformation can be only locally diffeomorphic, but the linearization results only hold in this smaller region.

We require several tools before we can solve this problem.

Lie Derivative

The goal of feedback linearization is to produce a transformed system whose states are the output y and its first $(n-1)$ derivatives. To understand the structure of this target system, we use the Lie derivative. Consider the time derivative of (2), which we can compute using the chain rule,

$$\dot{y} = \frac{\mathrm{d}\,h(x)}{\mathrm{d}\,t} = \frac{\mathrm{d}\,h(x)}{\mathrm{d}\,x}\dot{x}$$

$$= \frac{\mathrm{d}\,h(x)}{\mathrm{d}\,x}f(x) + \frac{\mathrm{d}\,h(x)}{\mathrm{d}\,x}g(x)u$$

Now we can define the Lie derivative of $h(x)$ along $f(x)$ as,

$$L_f h(x) = \frac{\mathrm{d}\,h(x)}{\mathrm{d}\,x}f(x),$$

and similarly, the Lie derivative of $h(x)$ along $g(x)$ as,

$$L_g h(x) = \frac{\mathrm{d}\,h(x)}{\mathrm{d}\,x}g(x).$$

With this new notation, we may express $\dot{y}$ as,

$$\dot{y} = L_f h(x) + L_g h(x)u$$

Note that the notation of Lie derivatives is convenient when we take multiple derivatives with respect to either the same vector field, or a different one. For example,

$$L_f^2 h(x) = L_f L_f h(x) = \frac{d(L_f h(x))}{dx} f(x),$$

and

$$L_g L_f h(x) = \frac{d(L_f h(x))}{dx} g(x).$$

Relative Degree

In our feedback linearized system made up of a state vector of the output y and its first $(n-1)$ derivatives, we must understand how the input u enters the system. To do this, we introduce the notion of relative degree. Our system given by (1) and (2) is said to have relative degree $r \in \mathbb{W}$ at a point x_0 if,

$$L_g L_f^k h(x) = 0 \qquad \forall x \text{ in a neighbourhood of } x_0 \text{ and all } k \leq r-2$$

$$L_g L_f^{r-1} h(x_0) \neq 0$$

Considering this definition of relative degree in light of the expression of the time derivative of the output y, we can consider the relative degree of our system (1) and (2) to be the number of times we have to differentiate the output y before the input u appears explicitly. In an LTI system, the relative degree is the difference between the degree of the transfer function's denominator polynomial (i.e., number of poles) and the degree of its numerator polynomial (i.e., number of zeros).

Linearization by Feedback

For the discussion that follows, we will assume that the relative degree of the system is n. In this case, after differentiating the output n times we have,

$$\begin{aligned}
y &= h(x) \\
\dot{y} &= L_f h(x) \\
\ddot{y} &= L_f^2 h(x) \\
&\vdots \\
y^{(n-1)} &= L_f^{n-1} h(x) \\
y^{(n)} &= L_f^n h(x) + L_g L_f^{n-1} h(x) u
\end{aligned}$$

where the notation $y^{(n)}$ indicates the nth derivative of y. Because we assumed the relative degree of the system is n, the Lie derivatives of the form $L_g L_f^i h(x)$ for $i = 1, \ldots, n-2$ are all zero. That is, the input u has no direct contribution to any of the first $(n-1)$th derivatives.

The coordinate transformation $T(x)$ that puts the system into normal form comes from the first $(n-1)$ derivatives. In particular,

$$z = T(x) = \begin{bmatrix} z_1(x) \\ z_2(x) \\ \vdots \\ z_n(x) \end{bmatrix} = \begin{bmatrix} y \\ \dot{y} \\ \vdots \\ y^{(n-1)} \end{bmatrix} = \begin{bmatrix} h(x) \\ L_f h(x) \\ \vdots \\ L_f^{n-1} h(x) \end{bmatrix}$$

transforms trajectories from the original x coordinate system into the new z coordinate system. So long as this transformation is a diffeomorphism, smooth trajectories in the original coordinate system will have unique counterparts in the z coordinate system that are also smooth. Those z trajectories will be described by the new system,

$$\begin{cases} \dot{z}_1 & = L_f h(x) = z_2(x) \\ \dot{z}_2 & = L_f^2 h(x) = z_3(x) \\ \quad \vdots \\ \dot{z}_n & = L_f^n h(x) + L_g L_f^{n-1} h(x) u \end{cases}$$

Hence, the feedback control law

$$u = \frac{1}{L_g L_f^{n-1} h(x)}(-L_f^n h(x) + v)$$

renders a linear input–output map from v to $z_1 = y$. The resulting linearized system

$$\begin{cases} \dot{z}_1 & = z_2 \\ \dot{z}_2 & = z_3 \\ \quad \vdots \\ \dot{z}_n & = v \end{cases}$$

is a cascade of n integrators, and an outer-loop control v may be chosen using standard linear system methodology. In particular, a state-feedback control law of

$$v = -Kz \quad ,$$

where the state vector z is the output y and its first $(n-1)$ derivatives, results in the LTI system

$$\dot{z} = Az$$

with,

$$A = \begin{bmatrix} 0 & 1 & 0 & \cdots & 0 \\ 0 & 0 & 1 & \cdots & 0 \\ \vdots & \vdots & \vdots & \ddots & \vdots \\ 0 & 0 & 0 & \cdots & 1 \\ -k_1 & -k_2 & -k_3 & \cdots & -k_n \end{bmatrix}.$$

So, with the appropriate choice of k, we can arbitrarily place the closed-loop poles of the linearized system.

Unstable Zero Dynamics

Feedback linearization can be accomplished with systems that have relative degree less than n. However, the normal form of the system will include zero dynamics (i.e., states that are not observable from the output of the system) that may be unstable. In practice, unstable dynamics may have deleterious effects on the system (e.g., it may be dangerous for internal states of the system to grow unbounded). These unobservable states may be controllable or at least stable, and so measures can be taken to ensure these states do not cause problems in practice.

Advanced Process Control

In control theory Advanced process control (APC) refers to a broad range of techniques and technologies implemented within industrial process control systems. Advanced process controls are usually deployed optionally and in addition to *basic* process controls. Basic process controls are designed and built with the process itself, to facilitate basic operation, control and automation requirements. Advanced process controls are typically added subsequently, often over the course of many years, to address particular performance or economic improvement opportunities in the process.

Process control (basic and advanced) normally implies the process industries, which includes chemicals, petrochemicals, oil and mineral refining, food processing, pharmaceuticals, power generation, etc. These industries are characterized by continuous processes and fluid processing, as opposed to discrete parts manufacturing, such as automobile and electronics manufacturing. The term process automation is essentially synonymous with process control.

Process controls (basic as well as advanced) are implemented within the process control system, which usually means a distributed control system (DCS), programmable logic controller (PLC), and/or a supervisory control computer. DCSs and PLCs are typically industrially hardened and fault-tolerant. Supervisory control computers are often not hardened or fault-tolerant, but they bring a higher level of computational capability to the control system, to host valuable, but not critical, advanced control applications. Advanced controls may reside in either the DCS or the supervisory computer, depending on the application. Basic controls reside in the DCS and its subsystems, including PLCs.

Types of Advanced Process Control

Following is a list of the best known types of advanced process control:

- Advanced regulatory control (ARC) refers to several proven advanced control techniques, such as feedforward, override or adaptive gain. ARC is also a catch-all term used to refer to any customized or non-simple technique that does not fall into any other category. ARCs are typically implemented using function blocks or custom programming capabilities at the DCS level. In some cases, ARCs reside at the supervisory control computer level.

- Multivariable Model predictive control (MPC) is a popular technology, usually deployed on a supervisory control computer, that identifies important independent and dependent process variables and the dynamic relationships (models) between them, and uses matrix-math based control and optimization algorithms, to control multiple variables simultaneously. One requirement of MPC is that the models must be linear (i.e. they must be repeatable at all points of the operating range of the controller). MPC has been a prominent part of APC ever since supervisory computers first brought the necessary computational capabilities to control systems in the 1980s.

- Inferential control: The concept behind inferentials is to calculate a stream property from readily available process measurements, such as temperature and pressure, that otherwise would require either an expensive and complicated online analyzer or periodic laboratory analysis. Inferentials can be utilized in place of actual online analyzers, whether for operator information, cascaded to base-layer process controllers, or multivariable controller CVs.

- Sequential control refers to dis-continuous time and event based automation sequences that occur within continuous processes. These may be implemented as a collection of time and logic function blocks, a custom algorithm, or using a formalized Sequential function chart methodology.

- Compressor control typically includes compressor anti-surge and performance control.

- Nonlinear MPC: Similar to Multivariable MPC in that it incorporates dynamic models and matrix-math based control; however, it does not have the requirement for model linearity. Nonlinear MPC is capable of accomodating processes with models that have varying process gains and dynamics (i.e. dead-times and lag times).

Related Technologies

The following technologies are related to APC and in some contexts can be considered part of APC, but are generally separate technologies having their own (or in need of their own).

- Statistical process control (SPC), despite its name, is much more common in discrete parts manufacturing and batch process control than in continuous process control. In SPC, "process" refers to the work and quality control process, rather than continuous process control.

- Batch process control is employed in non-continuous batch processes, such as many pharmaceuticals, chemicals, and foods.

- Simulation-based optimization incorporates dynamic or steady-state computer-based process simulation models to determine more optimal operating targets in real-time, i.e. on a periodic basis, ranging from hourly to daily. This is sometimes considered a part of APC, but in practice it is still an emerging technology and is more often part of MPO.

- Manufacturing planning and optimization (MPO) refers to ongoing business activity to arrive at optimal operating targets that are then implemented in the operating organization, either manually or in some cases automatically communicated to the process control system.

- Safety instrumented system refers to a system that is independent of the process control system, both physically and administratively, whose purpose is to assure basic safety of the process.

APC Business and Professionals

Those responsible for the design, implementation and maintenance of APC applications are often referred to as APC Engineers or Control Application Engineers. Usually their education is dependent upon the field of specialization. For example, in the process industries many APC Engineers have a chemical engineering background, combining process control and chemical processing expertise.

Most large operating facilities, such as oil refineries, employ a number of control system specialists and professionals, ranging from field instrumentation, regulatory control system (DCS and PLC), advanced process control, and control system network and security. Depending on facility size and circumstances, these personnel may have responsibilities across multiple areas, or be dedicated to each area. There are also many process control service companies that can be hired for support and services in each area.

Terminology

- APC: Advanced process control

- ARC: Advanced regulatory control, including feedforward, adaptive gain, override, logic, fuzzy logic, sequence control, device control, inferentials, and custom algorithms; usually implies DCS-based

- Base-Layer: Includes DCS, SIS, field devices, and other DCS subsystems, such as analyzers, equipment health systems, and PLCs

- BPCS: Basic process control system

- DCS: Distributed control system, often synonymous with BPCS

- MPO: Manufacturing planning optimization

- MPC: Multivariable Model predictive control

- SIS: Safety instrumented system

- SME: Subject matter expert

Automatic Control

Minimum human intervention is required to control many large
facilities such as this electrical generating station.

Automatic control is the application of control theory for regulation of processes without direct human intervention. In the simplest type of an automatic control loop, a controller compares a measured value of a process with a desired set value, and processes the resulting error signal to change some input to the process, in such a way that the process stays at its set point despite disturbances. This closed-loop control is an application of negative feedback to a system. The mathematical basis of control theory was begun in the 18th century, and advanced rapidly in the 20th.

Designing a system with features of automatic control generally requires the feeding of electrical or mechanical energy to enhance the dynamic features of an otherwise sluggish or variant, even errant system. The control is applied by regulating the energy feed.

Examples

Automatic control can self-regulate a technical plant (such as a machine or an industrial process) operating condition or parameters by the controller with minimal human intervention. A regulator such as a thermostat is an example of a device studied in automatic control. Another possible example of Automatic Control are the ABS of a car.

History of Automatic Control

Ancient Greece

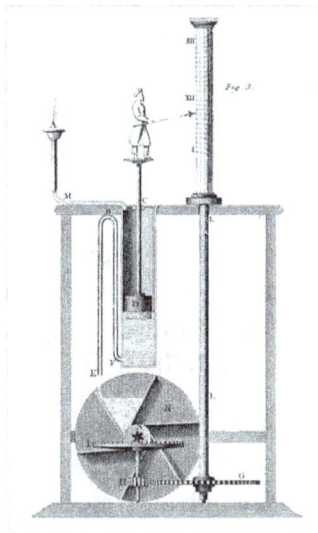

Ctesibius's clepsydra (3rd century BC).

It was a preoccupation of the Greeks and Arabs (in the period between about 300 BC and about 1200 AD) to keep accurate track of time. In about 270 BC the Greek Ctesibius invented a float regulator for a water clock, a device not unlike the ball and cock in a modern flush toilet. The invention of the mechanical clock in the 14th century made the water clock and its feedback control system obsolete. The float regulator does not appear again until its use in the Industrial Revolution.

Industrial Revolution in Europe

Thomas Newcomen invented the steam engine in 1713, and this date marks the accepted beginning of the Industrial Revolution; however, its roots can be traced back into the 17th century. The introduction of prime movers, or self-driven machines advanced grain mills, furnaces, boilers, and the steam engine created a new requirement for automatic control systems including temperature regulators (invented in 1624), pressure regulators (1681), float regulators (1700) and speed control devices. The design of feedback control systems up through the Indus-trial Revolution was by trial-and-error, together with a great deal of engineering intuition. Thus, it was more of an art than a science. In the mid-19th century mathematics was first used to analyze the stability of feedback control systems. Since mathematics is the formal language of automatic control theory, we could call the period before this time the prehistory of control theory.

First and Second World Wars

The First and Second World Wars saw major advancements in the field of mass communication and signal processing. Other key advances in automatic controls include differential equations, stability theory and system theory (1938), frequency domain analysis (1940), ship control (1950), and stochastic analysis (1941).

Space/Computer Age

With the advent of the space age in 1957, controls design, particularly in the United States, turned away from the frequency-domain techniques of classical control theory and backed into the dif-ferential equation techniques of the late 19th century, which were couched in the time domain. The modern era saw time-domain design for nonlinear systems (1961), navigation (1960), optimal control and estimation theory (1962), nonlinear control theory (1969), digital control and filtering theory (1974), and the personal computer (1983).

Robust Control

Robust control is a branch of control theory whose approach to controller design explicitly deals with uncertainty. Robust control methods are designed to function properly provided that uncer-tain parameters or disturbances are found within some (typically compact) set. Robust methods aim to achieve robust performance and/or stability in the presence of bounded modeling errors.

The early methods of Bode and others were fairly robust; the state-space methods invented in the 1960s and 1970s were sometimes found to lack robustness, prompting research to improve them. This was the start of the theory of Robust Control, which took shape in the 1980s and 1990s and is still active today.

In contrast with an adaptive control policy, a robust control policy is static; rather than adapting to measurements of variations, the controller is designed to work assuming that certain variables will be unknown but bounded.

When is a Control Method Said to be Robust?

Informally, a controller designed for a particular set of parameters is said to be robust if it also works well under a different set of assumptions. High-gain feedback is a simple example of a ro-bust control method; with sufficiently high gain, the effect of any parameter variations will be negligible. From the closed loop transfer function perspective, high open loop gain leads to substantial disturbance rejection in the face of system parameter uncertainty.

The major obstacle to achieving high loop gains is the need to maintain system closed loop stability. Loop shaping which allows stable closed loop operation can be a technical challenge.

Robust control systems often incorporate advanced topologies which include multiple feedback loops and feed-forward paths. The control laws may be represented by high order transfer functions required to simultaneously accomplish desired disturbance rejection performance with robust closed loop operation.

High-gain feedback is the principle that allows simplified models of operational amplifiers and emitter-degenerated bipolar transistors to be used in a variety of different settings. This idea was already well understood by Bode and Black in 1927.

The Modern Theory of Robust Control

The theory of robust control began in the late 1970s and early 1980s and soon developed a number of techniques for dealing with bounded system uncertainty.

Probably the most important example of a robust control technique is H-infinity loop-shaping, which was developed by Duncan McFarlane and Keith Glover of Cambridge University; this method minimizes the sensitivity of a system over its frequency spectrum, and this guarantees that the system will not greatly deviate from expected trajectories when disturbances enter the system.

An emerging area of robust control from application point of view is Sliding Mode Control (SMC), which is a variation of variable structure control (VSS). The robustness properties of SMC with respect to matched uncertainty as well as the simplicity in design attracted a variety of applications.

Another example is loop transfer recovery (LQG/LTR), which was developed to overcome the robustness problems of LQG control.

Other robust techniques includes Quantitative Feedback Theory (QFT), passivity based control, Lyapunov based controllers etc.

When system behavior varies considerably in normal operation, multiple control laws may have to be devised. Each distinct control law addresses a specific system behavior mode. An example is a computer hard disk drive. Separate robust control system modes are designed in order to address the rapid magnetic head traversal operation, known as the seek, a transitional settle operation as the magnetic head approaches its destination, and a track following mode during which the disk drive performs its data access operation.

One of the challenges is to design a control system that addresses these diverse system operating modes and enables smooth transition from one mode to the next as quickly as possible.

Such state machine driven composite control system is an extension of the gain scheduling idea where the entire control strategy changes based upon changes in system behavior.

Digital Control

Digital control is a branch of control theory that uses digital computers to act as system controllers. Depending on the requirements, a digital control system can take the form of a microcontroller to an ASIC to a standard desktop computer. Since a digital computer is a discrete system, the Laplace transform is replaced with the Z-transform. Also since a digital computer has finite precision, extra care is needed to ensure the error in coefficients, A/D conversion, D/A conversion, etc. are not producing undesired or unplanned effects.

The application of digital control can readily be understood in the use of feedback. Since the creation of the first digital computer in the early 1940s the price of digital computers has dropped considerably, which has made them key pieces to control systems for several reasons:

- Inexpensive: under $5 for many microcontrollers

- Flexible: easy to configure and reconfigure through software

- Scalable: programs can scale to the limits of the memory or storage space without extra cost

- Adaptable: parameters of the program can change with time

- Static operation: digital computers are much less prone to environmental conditions than capacitors, inductors, etc.

Digital Controller Implementation

A digital controller is usually cascaded with the plant in a feedback system. The rest of the system can either be digital or analog.

Typically, a digital controller requires:

- A/D conversion to convert analog inputs to machine readable (digital) format

- D/A conversion to convert digital outputs to a form that can be input to a plant (analog)

- A program that relates the outputs to the inputs

Output Program

- Outputs from the digital controller are functions of current and past input samples, as well as past output samples - this can be implemented by storing relevant values of input and output in registers. The output can then be formed by a weighted sum of these stored values.

The programs can take numerous forms and perform many functions

- A digital filter for low-pass filtering

- A state space model of a system to act as a state observer

- A telemetry system

Stability

Although a controller may be stable when implemented as an analog controller, it could be unstable when implemented as a digital controller due to a large sampling interval. During sampling the aliasing modifies the cutoff parameters. Thus the sample rate characterizes the transient response and stability of the compensated system, and must update the values at the controller input often enough so as to not cause instability.

When substituting the frequency into the z operator, regular stability criteria still apply to discrete control systems. Nyquist criteria apply to z-domain transfer functions as well as being general for complex valued functions. Bode stability criteria apply similarly. Jury criterion determines the discrete system stability about its characteristic polynomial.

Design of Digital Controller in S-domain

The digital controller can also be designed in the s-domain (continuous). The Tustin transformation can transform the continuous compensator to the respective digital compensator. The digital compensator will achieve an output which approaches the output of its respective analog controller as the sampling interval is decreased.

$$s = \frac{2(z-1)}{T(z+1)}$$

Tustin Transformation Deduction

Tustin is the Padé$_{(1,1)}$ approximation of the exponential function $z = e^{sT}$:

$$z = e^{sT}$$

$$= \frac{e^{sT/2}}{e^{-sT/2}}$$

$$\approx \frac{1 + sT/2}{1 - sT/2}$$

And its inverse

$$s = \frac{1}{T}\ln(z)$$

$$= \frac{2}{T}\left[\frac{z-1}{z+1} + \frac{1}{3}\left(\frac{z-1}{z+1}\right)^3 + \frac{1}{5}\left(\frac{z-1}{z+1}\right)^5 + \frac{1}{7}\left(\frac{z-1}{z+1}\right)^7 + \cdots\right]$$

$$\approx \frac{2}{T}\frac{z-1}{z+1}$$

$$= \frac{2}{T}\frac{1-z^{-1}}{1+z^{-1}}$$

We must never forget that the digital control theory is the technique to design strategies in discrete time, (and/or) quantized amplitude (and/or) in (binary) coded form to be implemented in computer systems (microcontrollers, microprocessors) that will control the analog (continuous in time and amplitude) dynamics of analog systems. From this consideration many errors from classical digital control were identified and solved and new methods were proposed:

- Marcelo Tredinnick and Marcelo Souza and their new type of analog-digital mapping

Systems Theory

Systems theory or systems science is the interdisciplinary study of systems in general, with the goal of discovering patterns and elucidating principles that can be discerned from and applied to all types of systems at all nesting levels in all fields of research. It can reasonably be considered a specialization of systems thinking or as the goal output of systems science and systems engineering, with an emphasis on generality useful across a broad range of systems (versus the particular models of individual fields).

A central topic of systems theory is self-regulating systems, self-correcting through feedback. Self-regulating systems are found in nature, including the physiological systems of our body, in local and global ecosystems, and in climate and also in human learning processes (from the individual on to international organizations like the UN).

Origin of the Term

The term originates from Bertalanffy's general system theory (GST) and is used in later efforts in other fields, such as the action theory of Talcott Parsons and the social systems theory of Niklas Luhmann.

Overview

Contemporary ideas from systems theory have grown with diverse areas, exemplified by the work of biologist Ludwig von Bertalanffy, linguist Béla H. Bánáthy, sociologist Talcott Parsons, ecological systems with Howard T. Odum, Eugene Odum and Fritjof Capra, organizational theory and management with individuals such as Peter Senge, interdisciplinary study with areas like Human Resource Development from the work of Richard A. Swanson, and insights from educators such as Debora Hammond and Alfonso Montuori. As a transdisciplinary, interdisciplinary and multiperspectival domain, the area brings together principles and concepts from ontology, philosophy of science, physics, computer science, biology, and engineering as well as geography, sociology, political science, psychotherapy (within family systems therapy) and economics among others. Systems theory thus serves as a bridge for interdisciplinary dialogue between autonomous areas of study as well as within the area of systems science itself.

In this respect, with the possibility of misinterpretations, von Bertalanffy believed a general theory of systems "should be an important regulative device in science," to guard against superficial analogies that "are useless in science and harmful in their practical consequences." Others remain closer to the direct systems concepts developed by the original theorists. For example, Ilya Prigogine, of the Center for Complex Quantum Systems at the University of Texas, Austin, has studied emergent properties, suggesting that they offer analogues for living systems. The theories of autopoiesis of Francisco Varela and Humberto Maturana represent further developments in this field. Important names in contemporary systems science include Russell Ackoff, Béla H. Bánáthy, Anthony Stafford Beer, Peter Checkland, Brian Wilson, Robert L. Flood, Fritjof Capra, Michael C. Jackson, and Edgar Morin among others.

With the modern foundations for a general theory of systems following World War I, Ervin Laszlo, in the preface for Bertalanffy's book: Perspectives on General System Theory, points out that the translation of "general system theory" from German into English has "wrought a certain amount of havoc":

It (General System Theory) was criticized as pseudoscience and said to be nothing more than an admonishment to attend to things in a holistic way. Such criticisms would have lost their point had it been recognized that von Bertalanffy's general system theory is a perspective or paradigm, and that such basic conceptual frameworks play a key role in the development of exact scientific theory. .. Allgemeine Systemtheorie is not directly consistent with an interpretation often put on 'general system theory,' to wit, that it is a (scientific) "theory of general systems." To criticize it as such is to shoot at straw men. Von Bertalanffy opened up something much broader and of much greater significance than a single theory (which, as we now know, can always be falsified and has usually an ephemeral existence): he created a new paradigm for the development of theories.".

"Theorie" (or "Lehre"), just as "Wissenschaft" (translated Scholarship), "has a much broader meaning in German than the closest English words 'theory' and 'science'". These ideas refer to an organized body of knowledge and "any systematically presented set of concepts, whether empirically, axiomatically, or philosophically" represented, while many associate "Lehre" with theory and science in the etymology of general systems, though it also does not translate from the German very well; its "closest equivalent" translates as "teaching", but "sounds dogmatic and off the mark". While the idea of a "general systems theory" might have lost many of its root meanings in the translation, by defining a new way of thinking about science and scientific paradigms, Systems theory became a widespread term used for instance to describe the interdependence of relationships created in organizations.

A system in this frame of reference can contain regularly interacting or interrelating groups of activities. For example, in noting the influence in organizational psychology as the field evolved from "an individually oriented industrial psychology to a systems and developmentally oriented organizational psychology", some theorists recognize that organizations have complex social systems; separating the parts from the whole reduces the overall effectiveness of organizations. This difference, from conventional models that center on individuals, structures, departments and units, separates in part from the whole, instead of recognizing the interdependence between groups of individuals, structures and processes that enable an organization to function. Laszlo explains that the new systems view of organized complexity went "one step beyond the Newtonian view of organized simplicity" which reduced the parts from the whole, or understood the whole without relation to the parts. The relationship between organisations and their environments can be seen as the foremost source of complexity and interdependence. In most cases, the whole has properties that cannot be known from analysis of the constituent elements in isolation. Béla H. Bánáthy, who argued — along with the founders of the systems society — that "the benefit of humankind" is the purpose of science, has made significant and far-reaching contributions to the area of systems theory. For the Primer Group at ISSS, Bánáthy defines a perspective that iterates this view:

The systems view is a world-view that is based on the discipline of SYSTEM INQUIRY. Central to systems inquiry is the concept of SYSTEM. In the most general sense, system means a configuration of parts connected and joined together by a web of relationships. The Primer group defines system as a family of relationships among the members acting as a whole. Von Bertalanffy defined system as "elements in standing relationship"

Similar ideas are found in learning theories that developed from the same fundamental concepts, emphasising how understanding results from knowing concepts both in part and as a whole. In fact, Bertalanffy's organismic psychology paralleled the learning theory of Jean Piaget. Some consider interdisciplinary perspectives critical in breaking away from industrial age models and thinking, wherein history represents history and math represents math, while the arts and sciences specialization remain separate and many treat teaching as behaviorist conditioning. The contemporary work of Peter Senge provides detailed discussion of the commonplace critique of educational systems grounded in conventional assumptions about learning, including the problems with fragmented knowledge and lack of holistic learning from the "machine-age thinking" that became a "model of school separated from daily life." In this way some systems theorists attempt to provide alternatives to, and evolved ideation from orthodox theories which have grounds in classical assumptions, including individuals such as Max Weber and Émile Durkheim in sociology and Frederick Winslow Taylor in scientific management. The theorists sought holistic methods by developing systems concepts that could integrate with different areas.

Some may view the contradiction of reductionism in conventional theory (which has as its subject a single part) as simply an example of changing assumptions. The emphasis with systems theory shifts from parts to the organization of parts, recognizing interactions of the parts as not static and constant but dynamic processes. Some questioned the conventional closed systems with the development of open systems perspectives. The shift originated from absolute and universal authoritative principles and knowledge to relative and general conceptual and perceptual knowledge and still remains in the tradition of theorists that sought to provide means to organize human life. In other words, theorists rethought the preceding history of ideas; they did not lose them. Mechanistic thinking was particularly critiqued, especially the industrial-age mechanistic metaphor for the mind from interpretations of Newtonian mechanics by Enlightenment philosophers and later psychologists that laid the foundations of modern organizational theory and management by the late 19th century.

Examples of Applications

Systems Biology

Systems biology is a movement that draws on several trends in bioscience research. Proponents describe systems biology as a biology-based inter-disciplinary study field that focuses on complex interactions in biological systems, claiming that it uses a new perspective (holism instead of reduction). Particularly from year 2000 onwards, the biosciences use the term widely and in a variety of contexts. An often stated ambition of systems biology is the modelling and discovery of emergent properties which represents properties of a system whose theoretical description requires the only possible useful techniques to fall under the remit of systems biology. It is thought that Ludwig von Bertalanffy may have created the term systems biology in 1928.

Systems Engineering

Systems engineering is an interdisciplinary approach and means for enabling the realisation and deployment of successful systems. It can be viewed as the application of engineering techniques to the engineering of systems, as well as the application of a systems approach to engineering efforts. Systems engineering integrates other disciplines and specialty groups into a team effort, forming a structured development process that proceeds from concept to production to operation and disposal. Systems engineering considers both the business and the technical needs of all customers, with the goal of providing a quality product that meets the user needs.

Systems Psychology

Systems psychology is a branch of psychology that studies human behaviour and experience in complex systems. It received inspiration from systems theory and systems thinking, as well as the basics of theoretical work from Roger Barker, Gregory Bateson, Humberto Maturana and others. It makes an approach in psychology in which groups and individuals receive consideration as systems in homeostasis. Systems psychology "includes the domain of engineering psychology, but in addition seems more concerned with societal systems and with the study of motivational, affective, cognitive and group behavior that holds the name engineering psychology." In systems psychology, "characteristics of organizational behaviour, for example individual needs, rewards, expectations, and attributes of the people interacting with the systems, considers this process in order to create an effective system".

History

Timeline

Precursors

- Saint-Simon (1760–1825), Karl Marx (1817–1883), Friedrich Engels (1820–1895), Herbert Spencer (1820–1903), Rudolf Clausius (1822–1888), Vilfredo Pareto (1848–1923), Émile Durkheim (1858–1917), Alexander Bogdanov (1873–1928), Nicolai Hartmann (1882–1950), Robert Maynard Hutchins (1929–1951), among others

Founders

- 1946-1953 Macy conferences

- 1948 Norbert Wiener publishes *Cybernetics or Control and Communication in the Animal and the Machine*

- 1951 Talcott Parsons publishes *The Social System*

- 1954 Ludwig von Bertalanffy, Anatol Rapoport, Ralph W. Gerard, Kenneth Boulding establish Society for the Advancement of General Systems Theory, in 1956 renamed to Society for General Systems Research.

- 1955 W. Ross Ashby publishes *Introduction to Cybernetics*

- 1968 Ludwig von Bertalanffy publishes *General System theory: Foundations, Development, Applications*

Other contributors

- 1970-1980s Second-order cybernetics developed by Heinz von Foerster, Gregory Bateson, Humberto Maturana and others

- 1971-1973 Cybersyn, rudimentary internet and cybernetic system for democratic economic planning developed in Chile under Allende government by Stafford Beer

- 1970s Catastrophe theory (René Thom, E.C. Zeeman) Dynamical systems in mathematics.

- 1977 Ilya Prigogine received the Nobel Prize for his works on self-organization, conciliating important *systems theory* concepts with system thermodynamics.

- 1980s Chaos theory, David Ruelle, Edward Lorenz, Mitchell Feigenbaum, Steve Smale, James A. Yorke

- 1986 Context theory, Anthony Wilden

- 1988 International Society for Systems Science

- 1990 Complex adaptive systems (CAS), John H. Holland, Murray Gell-Mann, W. Brian Arthur

Whether considering the first systems of written communication with Sumerian cuneiform to Mayan numerals, or the feats of engineering with the Egyptian pyramids, systems thinking can date back to antiquity. Differentiated from Western rationalist traditions of philosophy, C. West Churchman often identified with the I Ching as a systems approach sharing a frame of reference similar to pre-Socratic philosophy and Heraclitus. Von Bertalanffy traced systems concepts to the philosophy of G.W. Leibniz and Nicholas of Cusa's *coincidentia oppositorum*. While modern systems can seem considerably more complicated, today's systems may embed themselves in history.

Figures like James Joule and Sadi Carnot represent an important step to introduce the *systems approach* into the (rationalist) hard sciences of the 19th century, also known as the energy transformation. Then, the thermodynamics of this century, by Rudolf Clausius, Josiah Gibbs and others, established the *system* reference model as a formal scientific object.

The Society for General Systems Research specifically catalyzed systems theory as an area of study, which developed following the World Wars from the work of Ludwig von Bertalanffy, Anatol Rapoport, Kenneth E. Boulding, William Ross Ashby, Margaret Mead, Gregory Bateson, C. West

Churchman and others in the 1950s, had specifically catalyzed by collaboration in. Cognizant of advances in science that questioned classical assumptions in the organizational sciences, Bertalanffy's idea to develop a theory of systems began as early as the interwar period, publishing "An Outline for General Systems Theory" in the *British Journal for the Philosophy of Science*, Vol 1, No. 2, by 1950. Where assumptions in Western science from Greek thought with Plato and Aristotle to Newton's *Principia* have historically influenced all areas from the hard to social sciences, the original theorists explored the implications of twentieth century advances in terms of systems.

People have studied subjects like complexity, self-organization, connectionism and adaptive systems in the 1940s and 1950s. In fields like cybernetics, researchers such as Norbert Wiener, William Ross Ashby, John von Neumann and Heinz von Foerster, examined complex systems mathematically. John von Neumann discovered cellular automata and self-reproducing systems, again with only pencil and paper. Aleksandr Lyapunov and Jules Henri Poincaré worked on the foundations of chaos theory without any computer at all. At the same time Howard T. Odum, known as a radiation ecologist, recognized that the study of general systems required a language that could depict energetics, thermodynamics and kinetics at any system scale. Odum developed a general system, or universal language, based on the circuit language of electronics, to fulfill this role, known as the Energy Systems Language. Between 1929-1951, Robert Maynard Hutchins at the University of Chicago had undertaken efforts to encourage innovation and interdisciplinary research in the social sciences, aided by the Ford Foundation with the interdisciplinary Division of the Social Sciences established in 1931. Numerous scholars had actively engaged in these ideas before (Tectology by Alexander Bogdanov, published in 1912-1917, is a remarkable example), but in 1937, von Bertalanffy presented the general theory of systems at a conference at the University of Chicago.

The systems view was based on several fundamental ideas. First, all phenomena can be viewed as a web of relationships among elements, or a system. Second, all systems, whether electrical, biological, or social, have common patterns, behaviors, and properties that the observer can analyze and use to develop greater insight into the behavior of complex phenomena and to move closer toward a unity of the sciences. System philosophy, methodology and application are complementary to this science. By 1956, theorists established the Society for General Systems Research, which they renamed the International Society for Systems Science in 1988. The Cold War affected the research project for systems theory in ways that sorely disappointed many of the seminal theorists. Some began to recognize that theories defined in association with systems theory had deviated from the initial General Systems Theory (GST) view. The economist Kenneth Boulding, an early researcher in systems theory, had concerns over the manipulation of systems concepts. Boulding concluded from the effects of the Cold War that abuses of power always prove consequential and that systems theory might address such issues. Since the end of the Cold War, a renewed interest in systems theory emerged, combined with efforts to strengthen an ethical view on the subject.

Developments

General Systems Research and Systems Inquiry

Many early systems theorists aimed at finding a general systems theory that could explain all systems in all fields of science. The term goes back to Bertalanffy's book titled *"General System theory: Foundations, Development, Applications"* from 1968. According to Von Bertalanffy, he developed the "allgemeine Systemlehre" (general systems theory) first via lectures beginning in 1937 and then via publications beginning in 1946.

Von Bertalanffy's objective was to bring together under one heading the organismic science that he had observed in his work as a biologist. His desire was to use the word *system* for those principles that are common to systems in general. In GST, he writes:

...there exist models, principles, and laws that apply to generalized systems or their subclasses, irrespective of their particular kind, the nature of their component elements, and the relationships or "forces" between them. It seems legitimate to ask for a theory, not of systems of a more or less special kind, but of universal principles applying to systems in general.

Ervin Laszlo in the preface of von Bertalanffy's book *Perspectives on General System Theory*:

Thus when von Bertalanffy spoke of Allgemeine Systemtheorie it was consistent with his view that he was proposing a new perspective, a new way of doing science. It was not directly consistent with an interpretation often put on "general system theory", to wit, that it is a (scientific) "theory of general systems." To criticize it as such is to shoot at straw men. Von Bertalanffy opened up something much broader and of much greater significance than a single theory (which, as we now know, can always be falsified and has usually an ephemeral existence): he created a new paradigm for the development of theories.

Ludwig von Bertalanffy outlines systems inquiry into three major domains: Philosophy, Science, and Technology. In his work with the Primer Group, Béla H. Bánáthy generalized the domains into four integratable domains of systemic inquiry:

Domain	Description
Philosophy	the ontology, epistemology, and axiology of systems;
Theory	a set of interrelated concepts and principles applying to all systems
Methodology	the set of models, strategies, methods, and tools that instrumentalize systems theory and philosophy
Application	the application and interaction of the domains

These operate in a recursive relationship, he explained. Integrating Philosophy and Theory as Knowledge, and Method and Application as action, Systems Inquiry then is knowledgeable action.

Cybernetics

Cybernetics is the study of the communication and control of regulatory feedback both in living and lifeless systems (organisms, organizations, machines), and in combinations of those. Its focus is how anything (digital, mechanical or biological) controls its behavior, processes information, reacts to information, and changes or can be changed to better accomplish those three primary tasks.

The terms "systems theory" and "cybernetics" have been widely used as synonyms. Some authors use the term *cybernetic* systems to denote a proper subset of the class of general systems, namely those systems that include feedback loops. However Gordon Pask's differences of eternal interacting actor loops (that produce finite products) makes general systems a proper subset of cybernetics. According to Jackson (2000), von Bertalanffy promoted an embryonic form of general system theory (GST) as early as the 1920s and 1930s but it was not until the early 1950s it became more widely known in scientific circles.

Threads of cybernetics began in the late 1800s that led toward the publishing of seminal works (e.g., Wiener's *Cybernetics* in 1948 and von Bertalanffy's *General Systems Theory* in 1968). Cybernetics arose more from engineering fields and GST from biology. If anything it appears that although the two probably mutually influenced each other, cybernetics had the greater influence. Von Bertalanffy (1969) specifically makes the point of distinguishing between the areas in noting the influence of cybernetics: "Systems theory is frequently identified with cybernetics and control theory. This again is incorrect. Cybernetics as the theory of control mechanisms in technology and nature is founded on the concepts of information and feedback, but as part of a general theory of systems;" then reiterates: "the model is of wide application but should not be identified with 'systems theory' in general", and that "warning is necessary against its incautious expansion to fields for which its concepts are not made." (17-23). Jackson (2000) also claims von Bertalanffy was informed by Alexander Bogdanov's three volume *Tectology* that was published in Russia between 1912 and 1917, and was translated into German in 1928. He also states it is clear to Gorelik (1975) that the "conceptual part" of general system theory (GST) had first been put in place by Bogdanov. The similar position is held by Mattessich (1978) and Capra (1996). Ludwig von Bertalanffy never even mentioned Bogdanov in his works, which Capra (1996) finds "surprising".

Cybernetics, catastrophe theory, chaos theory and complexity theory have the common goal to explain complex systems that consist of a large number of mutually interacting and interrelated parts in terms of those interactions. Cellular automata (CA), neural networks (NN), artificial intelligence (AI), and artificial life (ALife) are related fields, but they do not try to describe general (universal) complex (singular) systems. The best context to compare the different "C"-Theories about complex systems is historical, which emphasizes different tools and methodologies, from pure mathematics in the beginning to pure computer science now. Since the beginning of chaos theory when Edward Lorenz accidentally discovered a strange attractor with his computer, computers have become an indispensable source of information. One could not imagine the study of complex systems without the use of computers today.

Complex Adaptive Systems

Complex adaptive systems (CAS) are special cases of complex systems. They are *complex* in that they are diverse and composed of multiple, interconnected elements; they are *adaptive* in that they have the capacity to change and learn from experience. In contrast to control systems in which negative feedback dampens and reverses disequilibria, CAS are often subject to positive feedback, which magnifies and perpetuates changes, converting local irregularities into global features. Another mechanism, Dual-phase evolution arises when connections between elements repeatedly change, shifting the system between phases of variation and selection that reshape the system.

The term *complex adaptive system* was coined at the interdisciplinary Santa Fe Institute (SFI), by

John H. Holland, Murray Gell-Mann and others. An alternative conception of complex adaptive (and learning) systems, methodologically at the interface between natural and social science, has been presented by Kristo Ivanov in terms of hypersystems. This concept intends to offer a theoretical basis for understanding and implementing participation of "users", decisions makers, designers and affected actors, in the development or maintenance of self-learning systems.

Control Systems: An Overview

This chapter gives an overview on control systems. A control system is a device that manages, commands or regulates the behavior of other systems or devices. Control systems can be applied to manual operations or machines that require or can facilitate an operator. The content on control systems offers an insightful focus, keeping in mind the complex subject matter.

A control system is a device, or set of devices, that manages, commands, directs or regulates the behaviour of other devices or systems. Industrial control systems are used in industrial production for controlling equipment or machines.

A hydroelectric power station in Amerongen, Netherlands.

There are two common classes of control systems, open loop control systems and closed loop control systems. In open loop control systems output is generated based on inputs. In closed loop control systems current output is taken into consideration and corrections are made based on feedback. A closed loop system is also called a feedback control system.

Overview

The term "control system" may be applied to the essentially manual controls that allow an operator, for example, to close and open a hydraulic press, perhaps including logic so that it cannot be moved unless safety guards are in place.

An automatic sequential control system may trigger a series of mechanical actuators in the correct sequence to perform a task. For example, various electric and pneumatic transducers may fold and glue a cardboard box, fill it with product and then seal it in an automatic packaging machine. Programmable logic controllers are used in many cases such as this, but several alternative technologies exist.

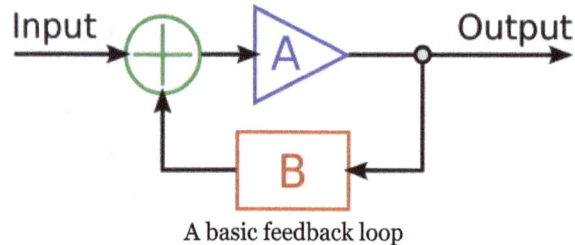

A basic feedback loop

In the case of linear feedback systems, a control loop, including sensors, control algorithms and actuators, is arranged in such a fashion as to try to regulate a variable at a setpoint or reference value. An example of this may increase the fuel supply to a furnace when a measured temperature drops. PID controllers are common and effective in cases such as this. Control systems that include some sensing of the results they are trying to achieve are making use of feedback and so can, to some extent, adapt to varying circumstances. Open-loop control systems do not make use of feedback, and run only in pre-arranged ways.

Logic Control

An internal lift control panel.

Logic control systems for industrial and commercial machinery were historically implemented at mains voltage using interconnected relays, designed using ladder logic. Today, most such systems are constructed with programmable logic controllers (PLCs) or microcontrollers. The notation of ladder logic is still in use as a programming idiom for PLCs.

Logic controllers may respond to switches, light sensors, pressure switches, etc., and can cause the machinery to start and stop various operations. Logic systems are used to sequence mechanical operations in many applications. PLC software can be written in many different ways – ladder diagrams, SFC – sequential function charts or in language terms known as statement lists.

Examples include elevators, washing machines and other systems with interrelated stop-go operations.

Logic systems are quite easy to design, and can handle very complex operations. Some aspects of logic system design make use of Boolean logic.

On–off Control

A thermostat is a simple negative feedback controller: when the temperature (the "process variable" or PV) goes below a set point (SP), the heater is switched on. Another example could be a pressure switch on an air compressor. When the pressure (PV) drops below the threshold (SP), the pump is powered. Refrigerators and vacuum pumps contain similar mechanisms operating in reverse, but still providing negative feedback to correct errors.

Simple on–off feedback control systems like these are cheap and effective. In some cases, like the simple compressor example, they may represent a good design choice.

In most applications of on–off feedback control, some consideration needs to be given to other costs, such as wear and tear of control valves and perhaps other start-up costs when power is reapplied each time the PV drops. Therefore, practical on–off control systems are designed to include hysteresis which acts as a deadband, a region around the setpoint value in which no control action occurs. The width of deadband may be adjustable or programmable.

Linear Control

Linear control systems use linear negative feedback to produce a control signal mathematically based on other variables, with a view to maintain the controlled process within an acceptable operating range.

The output from a linear control system into the controlled process may be in the form of a directly variable signal, such as a valve that may be 0 or 100% open or anywhere in between. Sometimes this is not feasible and so, after calculating the current required corrective signal, a linear control system may repeatedly switch an actuator, such as a pump, motor or heater, fully on and then fully off again, regulating the duty cycle using pulse-width modulation.

Proportional Control

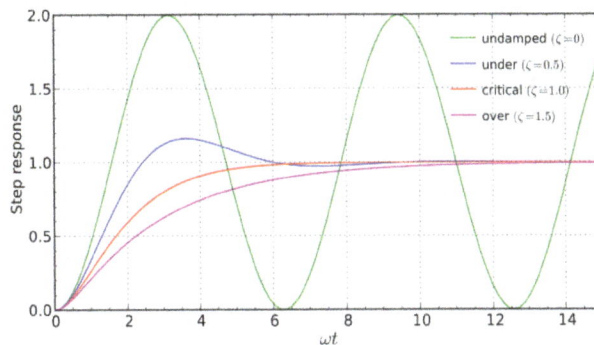

Step responses for a second order system defined by the transfer function

$$H(s) = \frac{\omega_n^2}{s^2 + 2\zeta\omega_n s + \omega_n^2},$$ where ζ is the damping ratio and ω_n is the undamped natural frequency.

When controlling the temperature of an industrial furnace, it is usually better to control the opening of the fuel valve *in proportion to* the current needs of the furnace. This helps avoid thermal shocks and applies heat more effectively.

Proportional negative-feedback systems are based on the difference between the required set point (SP) and process value (PV). This difference is called the *error*. Power is applied in direct proportion to the current measured error, in the correct sense so as to tend to reduce the error and therefore avoid positive feedback. The amount of corrective action that is applied for a given error is set by the gain or sensitivity of the control system.

At low gains, only a small corrective action is applied when errors are detected. The system may be safe and stable, but may be sluggish in response to changing conditions. Errors will remain uncorrected for relatively long periods of time and the system is over-damped. If the proportional gain is increased, such systems become more responsive and errors are dealt with more quickly. There is an optimal value for the gain setting when the overall system is said to be critically damped. Increases in loop gain beyond this point lead to oscillations in the PV and such a system is under-damped.

In real systems, there are practical limits to the range of the manipulated variable (MV). For example, a heater can be off or fully on, or a valve can be closed or fully open. Adjustments to the gain simultaneously alter the range of error values over which the MV is between these limits. The width of this range, in units of the error variable and therefore of the PV, is called the *proportional band* (PB). While the gain is useful in mathematical treatments, the proportional band is often used in practical situations. They both refer to the same thing, but the PB has an inverse relationship to gain – higher gains result in narrower PBs, and *vice versa*.

Under-damped Furnace Example

In the furnace example, suppose the temperature is increasing towards a set point at which, say, 50% of the available power will be required for steady-state. At low temperatures, 100% of available power is applied. When the process value (PV) is within, say 10° of the SP the heat input begins to be reduced by the proportional controller (note that this implies a 20° proportional band (PB) from full to no power input, evenly spread around the setpoint value). At the setpoint the controller will be applying 50% power as required, but stray stored heat within the heater sub-system and in the walls of the furnace will keep the measured temperature rising beyond what is required. At 10° above SP, we reach the top of the proportional band (PB) and no power is applied, but the temperature may continue to rise even further before beginning to fall back. Eventually as the PV falls back into the PB, heat is applied again, but now the heater and the furnace walls are too cool and the temperature falls too low before its fall is arrested, so that the oscillations continue.

Over-damped Furnace Example

The temperature oscillations that an under-damped furnace control system produces are unacceptable for many reasons, including the waste of fuel and time (each oscillation cycle may take many minutes), as well as the likelihood of seriously overheating both the furnace and its contents.

Suppose that the gain of the control system is reduced drastically and it is restarted. As the temperature approaches, say 30° below SP (60° proportional band (PB)), the heat input begins to be reduced, the rate of heating of the furnace has time to slow and, as the heat is still further reduced, it eventually is brought up to set point, just as 50% power input is reached and the furnace is operating as required. There was some wasted time while the furnace crept to its final temperature using only 52% then 51% of available power, but at least no harm was done. By carefully increasing the gain (i.e. reducing the width of the PB) this over-damped and sluggish behavior can be improved until the system is critically damped for this SP temperature. Doing this is known as 'tuning' the control system. A well-tuned proportional furnace temperature control system will usually be more effective than on-off control, but will still respond more slowly than the furnace could under skillful manual control.

PID Control

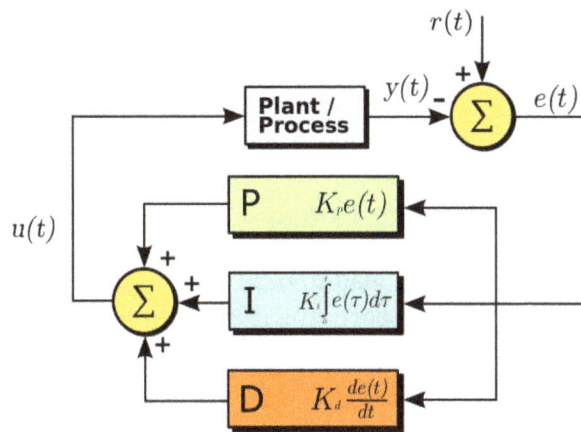

A block diagram of a PID controller

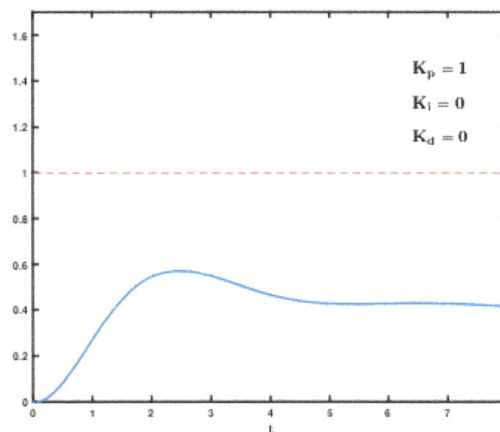

Effects of varying PID parameters (K_p, K_i, K_d) on the step response of a system.

Apart from sluggish performance to avoid oscillations, another problem with proportional-only control is that power application is always in direct proportion to the error. In the example above we assumed that the set temperature could be maintained with 50% power. What happens if the furnace is required in a different application where a higher set temperature will require 80% power to maintain it? If the gain was finally set to a 50° PB, then 80% power will not be applied unless the furnace is 15° below setpoint, so for this other application the operators will have to remember always to set the setpoint temperature 15° higher than actually needed. This 15° figure is not completely constant either: it will depend on the surrounding ambient temperature, as well as other factors that affect heat loss from or absorption within the furnace.

To resolve these two problems, many feedback control schemes include mathematical extensions to improve performance. The most common extensions lead to proportional-integral-derivative control, or PID control.

Derivative Action

The derivative part is concerned with the rate-of-change of the error with time: If the measured variable approaches the setpoint rapidly, then the actuator is backed off early to allow it to coast to the required level; conversely if the measured value begins to move rapidly away from the setpoint, extra effort is applied—in proportion to that rapidity—to try to maintain it.

Derivative action makes a control system behave much more intelligently. On control systems like the tuning of the temperature of a furnace, or perhaps the motion-control of a heavy item like a gun or camera on a moving vehicle, the derivative action of a well-tuned PID controller can allow it to reach and maintain a setpoint better than most skilled human operators could.

If derivative action is over-applied, it can lead to oscillations too. An example would be a PV that increased rapidly towards SP, then halted early and seemed to "shy away" from the setpoint before rising towards it again.

Integral Action

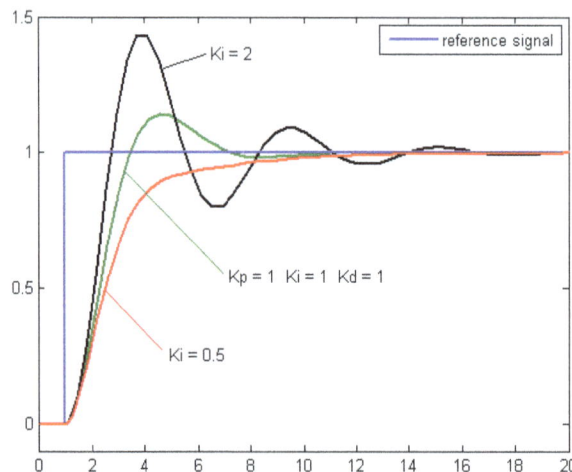

Change of response of second order system to a step input for varying Ki values.

The integral term magnifies the effect of long-term steady-state errors, applying ever-increasing effort until they reduce to zero. In the example of the furnace above working at various temperatures, if the heat being applied does not bring the furnace up to setpoint, for whatever reason, integral action increasingly *moves* the proportional band relative to the setpoint until the PV error is reduced to zero and the setpoint is achieved.

Ramp UP % Per minute

Some controllers include the option to limit the "ramp up % per minute". This option can be very helpful in stabilizing small boilers (3 MBTUH), especially during the summer, during light loads. A utility boiler "unit may be required to change load at a rate of as much as 5% per minute (IEA Coal Online - 2, 2007)".

Other Techniques

It is possible to filter the PV or error signal. Doing so can reduce the response of the system to undesirable frequencies, to help reduce instability or oscillations. Some feedback systems will oscillate at just one frequency. By filtering out that frequency, more "stiff" feedback can be applied, making the system more responsive without shaking itself apart.

Feedback systems can be combined. In cascade control, one control loop applies control algorithms to a measured variable against a setpoint, but then provides a varying setpoint to another control loop rather than affecting process variables directly. If a system has several different measured variables to be controlled, separate control systems will be present for each of them.

Control engineering in many applications produces control systems that are more complex than PID control. Examples of such fields include fly-by-wire aircraft control systems, chemical plants, and oil refineries. Model predictive control systems are designed using specialized computer-aided-design software and empirical mathematical models of the system to be controlled.

Fuzzy Logic

Fuzzy logic is an attempt to apply the easy design of logic controllers to the control of complex continuously varying systems. Basically, a measurement in a fuzzy logic system can be partly true, that is if yes is 1 and no is 0, a fuzzy measurement can be between 0 and 1.

The rules of the system are written in natural language and translated into fuzzy logic. For example, the design for a furnace would start with: "If the temperature is too high, reduce the fuel to the furnace. If the temperature is too low, increase the fuel to the furnace."

Measurements from the real world (such as the temperature of a furnace) are converted to values between 0 and 1 by seeing where they fall on a triangle. Usually, the tip of the triangle is the maximum possible value which translates to 1.

Fuzzy logic, then, modifies Boolean logic to be arithmetical. Usually the "not" operation is "output = 1 - input," the "and" operation is "output = input.1 multiplied by input.2," and "or" is "output = 1 - ((1 - input.1) multiplied by (1 - input.2))". This reduces to Boolean arithmetic if values are restricted to 0 and 1, instead of allowed to range in the unit interval [0,1].

The last step is to "defuzzify" an output. Basically, the fuzzy calculations make a value between zero and one. That number is used to select a value on a line whose slope and height converts the fuzzy value to a real-world output number. The number then controls real machinery.

If the triangles are defined correctly and rules are right the result can be a good control system.

When a robust fuzzy design is reduced into a single, quick calculation, it begins to resemble a conventional feedback loop solution and it might appear that the fuzzy design was unnecessary. However, the fuzzy logic paradigm may provide scalability for large control systems where conventional methods become unwieldy or costly to derive.

Fuzzy electronics is an electronic technology that uses fuzzy logic instead of the two-value logic more commonly used in digital electronics.

Physical Implementations

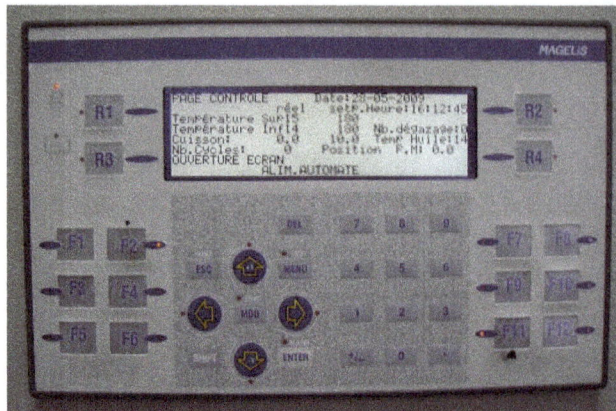

A control panel of a hydraulic heat press machine.

Since modern small microprocessors are so cheap (often less than $1 US), it's very common to implement control systems, including feedback loops, with computers, often in an embedded system. The feedback controls are simulated by having the computer make periodic measurements and then calculate from this stream of measurements.

Computers emulate logic devices by making measurements of switch inputs, calculating a logic function from these measurements and then sending the results out to electronically controlled switches.

Logic systems and feedback controllers are usually implemented with programmable logic controllers which are devices available from electrical supply houses. They include a little computer and a simplified system for programming. Most often they are programmed with personal computers.

Logic controllers have also been constructed from relays, hydraulic and pneumatic devices as well as electronics using both transistors and vacuum tubes (feedback controllers can also be constructed in this manner).

Types of Control Systems

Control systems can best be understood in confluence with the major topics listed in the following chapter. The major categories of control systems are dealt with great detail in the chapter. Industrial control system, PID controller, fly by wire are some of the control systems analyzed in this chapter. The topics discussed are of great importance to broaden the existing knowledge on control systems.

Industrial Control System

Industrial control system (ICS) is a general term that encompasses several types of control systems used in industrial production, including supervisory control and data acquisition (SCADA) systems, distributed control systems (DCS), and other smaller control system configurations such as programmable logic controllers (PLC) often found in the industrial sectors and critical infrastructures.

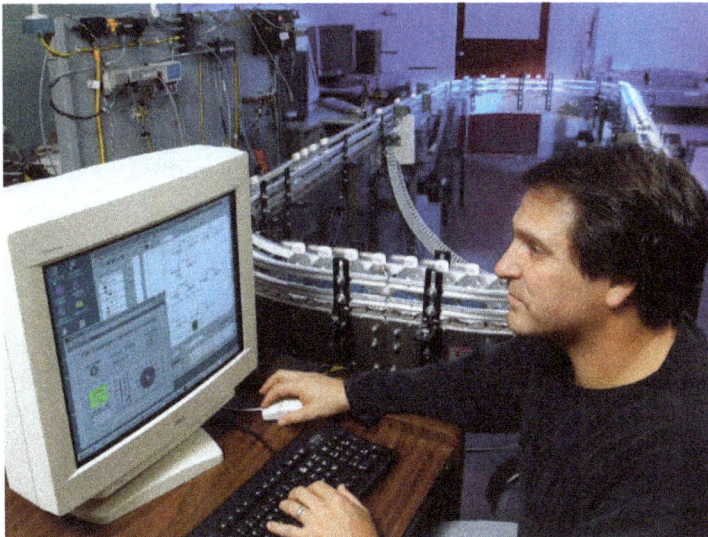

NIST Industrial Control Security Testbed.

ICSs are typically used in industries such as electrical, water, oil, gas and data. Based on data received from remote stations, automated or operator-driven supervisory commands can be pushed to remote station control devices, which are often referred to as field devices. Field devices control local operations such as opening and closing valves and breakers, collecting data from sensor systems, and monitoring the local environment for alarm conditions.

A historical Perspective

Industrial control system technology has evolved over the decades.

DCS (distributed control systems) generally refer to the particular functional distributed control system design that exist in industrial process plants (e.g., oil and gas, refining, chemical, pharmaceutical, some food and beverage, water and wastewater, pulp and paper, utility power, mining, metals). The DCS concept came about from a need to gather data and control the systems on a large campus in real time on high-bandwidth, low-latency data networks. It is common for loop controls to extend all the way to the top level controllers in a DCS, as everything works in real time. These systems evolved from a need to extend pneumatic control systems beyond just a small cell area of a refinery.

PLC (programmable logic controller) evolved out of a need to replace racks of relays in ladder form. The latter were not particularly reliable, were difficult to rewire, and were difficult to diagnose. PLC control tends to be used in very regular, high-speed binary controls, such as controlling a high-speed printing press. Originally, PLC equipment did not have remote I/O racks, and many could not perform more than rudimentary analog controls.

SCADA's history is rooted in distribution applications, such as power, natural gas, and water pipelines, where there is a need to gather remote data through potentially unreliable or intermittent low-bandwidth/high-latency links. SCADA systems use open-loop control with sites that are widely separated geographically. A SCADA system uses RTUs (remote terminal units, also referred to as remote telemetry units) to send supervisory data back to a control center. Most RTU systems always did have some limited capacity to handle local controls while the master station is not available. However, over the years RTU systems have grown more and more capable of handling local controls.

The boundaries between these system definitions are blurring as time goes on. The technical limits that drove the designs of these various systems are no longer as much of an issue. Many PLC platforms can now perform quite well as a small DCS, using remote I/O and are sufficiently reliable that some SCADA systems actually manage closed loop control over long distances. With the increasing speed of today's processors, many DCS products have a full line of PLC-like subsystems that weren't offered when they were initially developed.

This led to the concept of a PAC (programmable automation controller or process automation controller), that is an amalgamation of these three concepts. Time and the market will determine whether this can simplify some of the terminology and confusion that surrounds these concepts today.

DCSs

DCSs are used to control industrial processes such as electric power generation, oil and gas refineries, water and wastewater treatment, and chemical, food, and automotive production. DCSs are integrated as a control architecture containing a supervisory level of control, overseeing multiple integrated sub-systems that are responsible for controlling the details of a localized process.

Product and process control are usually achieved by deploying feed back or feed forward control loops whereby key product and/or process conditions are automatically maintained around a desired set point. To accomplish the desired product and/or process tolerance around a specified set point, only specific programmable controllers are used.

PLCs

PLCs provide boolean logic operations, timers, and (in some models) continuous control. The proportional, integral, and/or differential gains of the PLC continuous control feature may be tuned to provide the desired tolerance as well as the rate of self-correction during process upsets. PLCs are used extensively in process-based industries. PLCs are computer-based solid-state devices that control industrial equipment and processes. While PLCs can control system components used throughout SCADA and DCS systems, they are often the primary components in smaller control system configurations. They are used to provide regulatory control of discrete processes such as automobile assembly lines and power plant soot blower controls and are used extensively in almost all industrial processes.

Embedded Control

Another option is the use of several small embedded controls attached to an industrial computer via a network. Examples are the Lantronix Xport and Digi/ME.

PID Controller

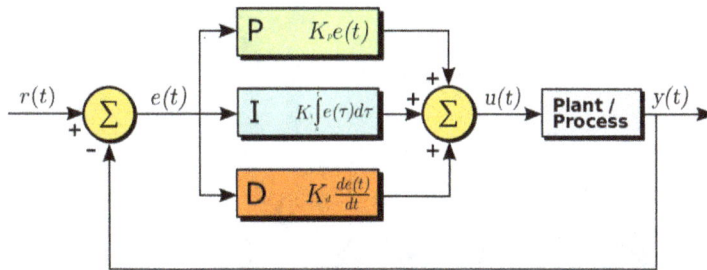

A block diagram of a PID controller in a feedback loop

A proportional–integral–derivative controller (PID controller) is a control loop feedback mechanism (controller) commonly used in industrial control systems. A PID controller continuously calculates an *error value* as the difference between a desired setpoint and a measured process variable. The controller attempts to minimize the error over time by adjustment of a *control variable*, such as the position of a control valve, a damper, or the power supplied to a heating element, to a new value determined by a weighted sum:

$$u(t) = K_p e(t) + K_i \int_0^t e(\tau)d\tau + K_d \frac{de(t)}{dt}$$

where K_p, K_i, and K_d, all non-negative, denote the coefficients for the proportional, integral, and derivative terms, respectively (sometimes denoted *P, I,* and *D*). In this model,

- *P* accounts for present values of the error. For example, if the error is large and positive, the control output will also be large and positive.

- *I* accounts for past values of the error. For example, if the current output is not sufficiently strong, error will accumulate over time, and the controller will respond by applying a stronger action.

- *D* accounts for possible future values of the error, based on its current rate of change.

As a PID controller relies only on the measured process variable, not on knowledge of the underlying process, it is broadly applicable. By tuning the three parameters of the model, a PID controller can deal with specific process requirements. The response of the controller can be described in terms of its responsiveness to an error, the degree to which the system overshoots a setpoint, and the degree of any system oscillation. The use of the PID algorithm does not guarantee optimal control of the system or even its stability.

Some applications may require using only one or two terms to provide the appropriate system control. This is achieved by setting the other parameters to zero. A PID controller will be called a PI, PD, P or I controller in the absence of the respective control actions. PI controllers are fairly common, since derivative action is sensitive to measurement noise, whereas the absence of an integral term may prevent the system from reaching its target value.

For discrete time systems, the term PSD, for proportional-summation-difference, is often used.

History and Applications

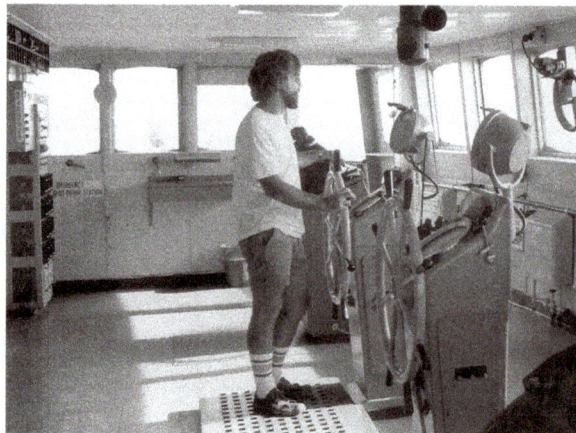

PID theory developed by observing the action of helmsmen.

PID controllers date to 1890s governor design. PID controllers were subsequently developed in automatic ship steering. One of the earliest examples of a PID-type controller was developed by Elmer Sperry in 1911, while the first published theoretical analysis of a PID controller was by Russian American engineer Nicolas Minorsky, (Minorsky 1922). Minorsky was designing automatic steering systems for the US Navy, and based his analysis on observations of a helmsman, noting the helmsman controlled the ship based not only on the current error, but also on past error as well as the current rate of change; this was then made mathematical by Minorsky. His goal was stability, not general control, which simplified the problem significantly. While proportional control provides stability against small disturbances, it was insufficient for dealing with a steady disturbance, notably a stiff gale (due to steady-state error), which required adding the integral term. Finally, the derivative term was added to improve stability and control.

Trials were carried out on the USS *New Mexico*, with the controller controlling the *angular velocity* (not angle) of the rudder. PI control yielded sustained yaw (angular error) of ±2°. Adding the D element yielded a yaw error of ±1/6°, better than most helmsmen could achieve.

The Navy ultimately did not adopt the system, due to resistance by personnel. Similar work was carried out and published by several others in the 1930s.

In the early history of automatic process control the PID controller was implemented as a mechanical device. These mechanical controllers used a lever, spring and a mass and were often energized by compressed air. These pneumatic controllers were once the industry standard.

Electronic analog controllers can be made from a solid-state or tube amplifier, a capacitor and a resistor. Electronic analog PID control loops were often found within more complex electronic systems, for example, the head positioning of a disk drive, the power conditioning of a power supply, or even the movement-detection circuit of a modern seismometer. Nowadays, electronic controllers have largely been replaced by digital controllers implemented with microcontrollers or FPGAs. However, analog PID controllers are still used in niche applications requiring high-bandwidth and low noise performance, such as laser diode controllers.

Most modern PID controllers in industry are implemented in programmable logic controllers (PLCs) or as a panel-mounted digital controller. Software implementations have the advantages that they are relatively cheap and are flexible with respect to the implementation of the PID algorithm. PID temperature controllers are applied in industrial ovens, plastics injection machinery, hot stamping machines and packing industry.

Control Loop Basics

A robotic arm can be moved and positioned by a control loop. By applying forward and reverse power to an electric motor to lift and lower the arm, it may be necessary to allow for the inertial mass of the arm, forces due to gravity, and to correct for external forces on the arm such as a load to lift or work to be done on an external object.

The sensed position is the process variable (PV). The desired position is called the setpoint (SP). The input to the process (the electric current in the motor) is the output from the PID controller. It is called either the manipulated variable (MV) or the control variable (CV). The difference between the present position and the setpoint is the error (e), which quantifies whether the arm is too low or too high and by how much.

By measuring the position (PV), and subtracting it from the setpoint (SP), the error (e) is found, and from it the controller calculates how much electric current to supply to the motor (MV). The obvious method is proportional control: the motor current is set in proportion to the existing error. A more complex control may include another term: derivative action. This considers the rate of change of error, supplying more or less electric current depending on how fast the error is approaching zero. Finally, integral action adds a third term, using the accumulated position error in the past to detect whether the position of the mechanical arm is settling out too low or too high and to set the electrical current in relation not only to the error but also the time for which it has persisted. An alternative formulation of integral action is to change the electric current in small persistent steps that are proportional to the current error.

Over time the steps accumulate and add up dependent on past errors; this is the discrete-time equivalent to integration.

Applying too much impetus when the error is small and is reducing will lead to overshoot. After overshooting, if the controller were to apply a large correction in the opposite direction and repeatedly overshoot the desired position, the output would oscillate around the setpoint in either a constant, growing, or decaying sinusoid. If the amplitude of the oscillations increase with time, the system is unstable. If they decrease, the system is stable. If the oscillations remain at a constant magnitude, the system is marginally stable.

In the interest of achieving a controlled arrival at the desired position (SP) in a timely and accurate way, the controlled system needs to be critically damped. A well-tuned position control system will also apply the necessary currents to the controlled motor so that the arm pushes and pulls as necessary to resist external forces trying to move it away from the required position. The setpoint itself may be generated by an external system, such as a PLC or other computer system, so that it continuously varies depending on the work that the robotic arm is expected to do. A well-tuned PID control system will enable the arm to meet these changing requirements to the best of its capabilities.

If a controller starts from a stable state with zero error (PV = SP), then further changes by the controller will be in response to changes in other measured or unmeasured inputs to the process that affect the process, and hence the PV. Variables that affect the process other than the MV are known as disturbances. Generally controllers are used to reject disturbances and to implement setpoint changes. A change in load on the arm constitutes a disturbance to the robot arm control process.

In theory, a controller can be used to control any process which has a measurable output (PV), a known ideal value for that output (SP) and an input to the process (MV) that will affect the relevant PV. Controllers are used in industry to regulate temperature, pressure, force, feed, flow rate, chemical composition, weight, position, speed and practically every other variable for which a measurement exists.

PID Controller Theory

The PID control scheme is named after its three correcting terms, whose sum constitutes the manipulated variable (MV). The proportional, integral, and derivative terms are summed to calculate the output of the PID controller. Defining $u(t)$ as the controller output, the final form of the PID algorithm is:

$$u(t) = \text{MV}(t) = K_p e(t) + K_i \int_0^t e(\tau)d\tau + K_d \frac{de(t)}{dt}$$

where

K_p : Proportional gain, a tuning parameter

K_i : Integral gain, a tuning parameter

K_d : Derivative gain, a tuning parameter

$e(t)$: Error $= SP - PV(t)$

SP: Set Point

$PV(t)$: Process Variable

t: Time or instantaneous time (the present)

τ: Variable of integration; takes on values from time 0 to the present t.

Equivalently, the transfer function in the Laplace Domain of the PID controller is

$$L(s) = K_p + K_i / s + K_d s$$

where

s: complex number frequency

Proportional Term

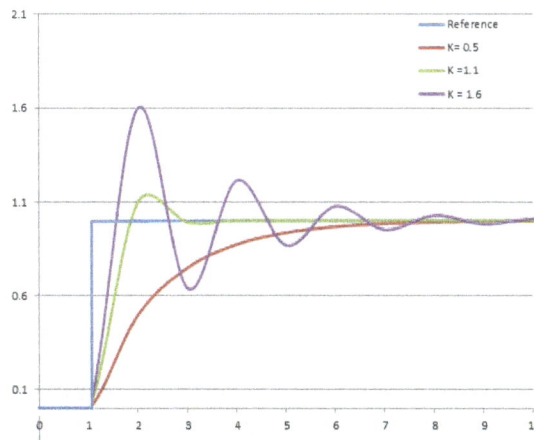

Plot of PV vs time, for three values of K_p (K_i and K_d held constant)

The proportional term produces an output value that is proportional to the current error value. The proportional response can be adjusted by multiplying the error by a constant K_p, called the proportional gain constant.

The proportional term is given by:

$$P_{out} = K_p\, e(t)$$

A high proportional gain results in a large change in the output for a given change in the error. If the proportional gain is too high, the system can become unstable. In contrast, a small gain results in a small output response to a large input error, and a less responsive or less sensitive controller. If the proportional gain is too low, the control action may be too small when responding to system disturbances. Tuning theory and industrial practice indicate that the proportional term should contribute the bulk of the output change.

Steady-state Error

Because a non-zero error is required to drive it, a proportional controller generally operates with a so-called *steady-state error*. Steady-state error (SSE) is proportional to the process gain and inversely proportional to proportional gain. SSE may be mitigated by adding a compensating bias term to the setpoint or output, or corrected dynamically by adding an integral term.

Integral Term

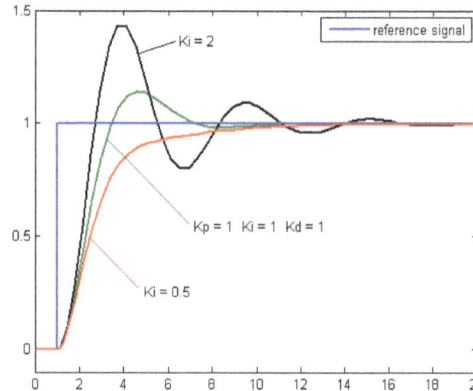

Plot of PV vs time, for three values of K_i (K_p and K_d held constant)

The contribution from the integral term is proportional to both the magnitude of the error and the duration of the error. The integral in a PID controller is the sum of the instantaneous error over time and gives the accumulated offset that should have been corrected previously. The accumulated error is then multiplied by the integral gain (K_i) and added to the controller output.

The integral term is given by:

$$I_{out} = K_i \int_0^t e(\tau) d\tau$$

The integral term accelerates the movement of the process towards setpoint and eliminates the residual steady-state error that occurs with a pure proportional controller. However, since the integral term responds to accumulated errors from the past, it can cause the present value to overshoot the setpoint value.

Derivative Term

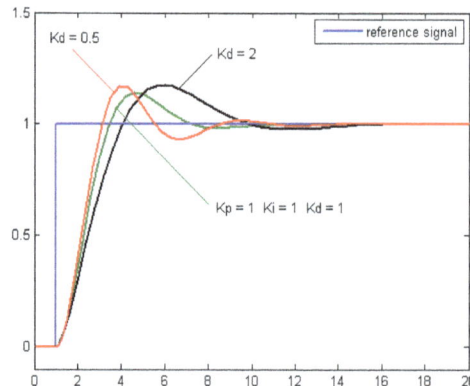

Plot of PV vs time, for three values of K_d (K_p and K_i held constant)

The derivative of the process error is calculated by determining the slope of the error over time and multiplying this rate of change by the derivative gain K_d. The magnitude of the contribution of the derivative term to the overall control action is termed the derivative gain, K_d.

The derivative term is given by:

$$D_{out} = K_d \frac{de(t)}{dt}$$

Derivative action predicts system behavior and thus improves settling time and stability of the system. An ideal derivative is not causal, so that implementations of PID controllers include an additional low pass filtering for the derivative term, to limit the high frequency gain and noise. Derivative action is seldom used in practice though - by one estimate in only 25% of deployed controllers - because of its variable impact on system stability in real-world applications.

Loop Tuning

Tuning a control loop is the adjustment of its control parameters (proportional band/gain, integral gain/reset, derivative gain/rate) to the optimum values for the desired control response. Stability (no unbounded oscillation) is a basic requirement, but beyond that, different systems have different behavior, different applications have different requirements, and requirements may conflict with one another.

PID tuning is a difficult problem, even though there are only three parameters and in principle is simple to describe, because it must satisfy complex criteria within the limitations of PID control. There are accordingly various methods for loop tuning, and more sophisticated techniques are the subject of patents; this section describes some traditional manual methods for loop tuning.

Designing and tuning a PID controller appears to be conceptually intuitive, but can be hard in practice, if multiple (and often conflicting) objectives such as short transient and high stability are to be achieved. PID controllers often provide acceptable control using default tunings, but performance can generally be improved by careful tuning, and performance may be unacceptable with poor tuning. Usually, initial designs need to be adjusted repeatedly through computer simulations until the closed-loop system performs or compromises as desired.

Some processes have a degree of nonlinearity and so parameters that work well at full-load conditions don't work when the process is starting up from no-load; this can be corrected by gain scheduling (using different parameters in different operating regions).

Stability

If the PID controller parameters (the gains of the proportional, integral and derivative terms) are chosen incorrectly, the controlled process input can be unstable, i.e., its output diverges, with or without oscillation, and is limited only by saturation or mechanical breakage. Instability is caused by *excess* gain, particularly in the presence of significant lag.

Generally, stabilization of response is required and the process must not oscillate for any combination of process conditions and setpoints, though sometimes marginal stability (bounded oscillation) is acceptable or desired.

Mathematically, the origins of instability can be seen in the Laplace domain. The total loop transfer function is:

$$H(s) = \frac{K(s)G(s)}{1 + K(s)G(s)}$$

where

$K(s)$: PID transfer function

$G(s)$: Plant transfer function

The system is called unstable where the closed loop transfer function diverges for some s. This happens for situations where $K(s)G(s) = -1$. Typically, this happens when $|K(s)G(s)| = 1$ with a 180 degree phase shift. Stability is guaranteed when $K(s)G(s) < 1$ for frequencies that suffer high phase shifts. A more general formalism of this effect is known as the Nyquist stability criterion.

Optimum Behavior

The optimum behavior on a process change or setpoint change varies depending on the application.

Two basic requirements are *regulation* (disturbance rejection – staying at a given setpoint) and *command tracking* (implementing setpoint changes) – these refer to how well the controlled variable tracks the desired value. Specific criteria for command tracking include rise time and settling time. Some processes must not allow an overshoot of the process variable beyond the setpoint if, for example, this would be unsafe. Other processes must minimize the energy expended in reaching a new setpoint.

Overview of Methods

There are several methods for tuning a PID loop. The most effective methods generally involve the development of some form of process model, then choosing P, I, and D based on the dynamic model parameters. Manual tuning methods can be relatively time consuming, particularly for systems with long loop times.

The choice of method will depend largely on whether or not the loop can be taken "offline" for tuning, and on the response time of the system. If the system can be taken offline, the best tuning method often involves subjecting the system to a step change in input, measuring the output as a function of time, and using this response to determine the control parameters.

Choosing a tuning method		
Method	**Advantages**	**Disadvantages**
Manual tuning	No math required; online.	Requires experienced personnel.
Ziegler–Nichols	Proven method; online.	Process upset, some trial-and-error, very aggressive tuning.
Tyreus Luyben	Proven method; online.	Process upset, some trial-and-error, very aggressive tuning.

Software tools	Consistent tuning; online or offline - can employ computer-automated control system design (*CAutoD*) techniques; may include valve and sensor analysis; allows simulation before downloading; can support non-steady-state (NSS) tuning.	Some cost or training involved.
Cohen–Coon	Good process models.	Some math; offline; only good for first-order processes.
Åström–Hägglund	Can be used for auto tuning; amplitude is minimum so this method has lowest process upset	The process itself is inherently oscillatory.

Manual Tuning

If the system must remain online, one tuning method is to first set K_i and K_d values to zero. Increase the K_p until the output of the loop oscillates, then the K_p should be set to approximately half of that value for a "quarter amplitude decay" type response. Then increase K_i until any offset is corrected in sufficient time for the process. However, too much K_i will cause instability. Finally, increase K_d, if required, until the loop is acceptably quick to reach its reference after a load disturbance. However, too much K_d will cause excessive response and overshoot. A fast PID loop tuning usually overshoots slightly to reach the setpoint more quickly; however, some systems cannot accept overshoot, in which case an *over-damped* closed-loop system is required, which will require a K_p setting significantly less than half that of the K_p setting that was causing oscillation.

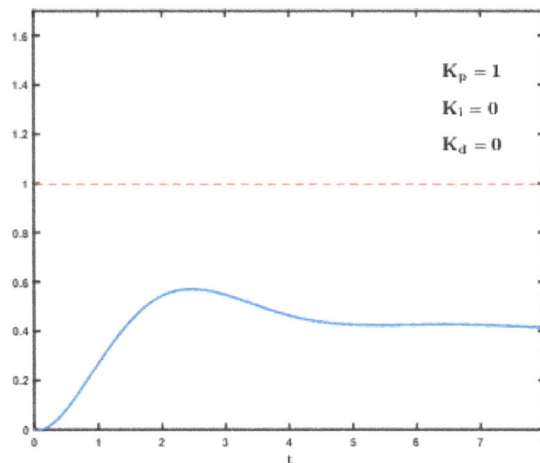

Effects of varying PID parameters (K_p, K_i, K_d) on the step response of a system.

Effects of *increasing* a parameter independently					
Parameter	Rise time	Overshoot	Settling time	Steady-state error	Stability
K_p	Decrease	Increase	Small change	Decrease	Degrade
K_i	Decrease	Increase	Increase	Eliminate	Degrade
K_d	Minor change	Decrease	Decrease	No effect in theory	Improve if K_d small

Ziegler–nichols Method

Another heuristic tuning method is formally known as the Ziegler–Nichols method, introduced by John G. Ziegler and Nathaniel B. Nichols in the 1940s. As in the method above, the K_i and K_d gains are first set to zero. The proportional gain is increased until it reaches the ultimate gain, K_u, at which the output of the loop starts to oscillate. K_u and the oscillation period T_u are used to set the gains as shown:

Ziegler–Nichols method			
Control Type	K_p	K_i	K_d
P	$0.50K_u$	-	-
PI	$0.45K_u$	$1.2K_p / T_u$	-
PID	$0.60K_u$	$2K_p / T_u$	$K_pT_u / 8$

These gains apply to the ideal, parallel form of the PID controller. When applied to the standard PID form, the integral and derivative time parameters T_i and T_d are only dependent on the oscillation period T_u.

PID Tuning Software

Most modern industrial facilities no longer tune loops using the manual calculation methods shown above. Instead, PID tuning and loop optimization software are used to ensure consistent results. These software packages will gather the data, develop process models, and suggest optimal tuning. Some software packages can even develop tuning by gathering data from reference changes.

Mathematical PID loop tuning induces an impulse in the system, and then uses the controlled system's frequency response to design the PID loop values. In loops with response times of several minutes, mathematical loop tuning is recommended, because trial and error can take days just to find a stable set of loop values. Optimal values are harder to find. Some digital loop controllers offer a self-tuning feature in which very small setpoint changes are sent to the process, allowing the controller itself to calculate optimal tuning values.

Other formulas are available to tune the loop according to different performance criteria. Many patented formulas are now embedded within PID tuning software and hardware modules.

Advances in automated PID Loop Tuning software also deliver algorithms for tuning PID Loops in a dynamic or Non-Steady State (NSS) scenario. The software will model the dynamics of a process, through a disturbance, and calculate PID control parameters in response.

Limitations of PID Control

While PID controllers are applicable to many control problems, and often perform satisfactorily without any improvements or only coarse tuning, they can perform poorly in some applications, and do not in general provide *optimal* control. The fundamental difficulty with PID control is that it is a feedback control system, with *constant* parameters, and no direct knowledge of the process, and thus overall performance is reactive and a compromise. While PID control is the best controller in an observer without a model of the process, better performance can be obtained by overtly modeling the actor of the process without resorting to an observer.

PID controllers, when used alone, can give poor performance when the PID loop gains must be reduced so that the control system does not overshoot, oscillate or hunt about the control setpoint value. They also have difficulties in the presence of non-linearities, may trade-off regulation versus response time, do not react to changing process behavior (say, the process changes after it has warmed up), and have lag in responding to large disturbances.

The most significant improvement is to incorporate feed-forward control with knowledge about the system, and using the PID only to control error. Alternatively, PIDs can be modified in more minor ways, such as by changing the parameters (either gain scheduling in different use cases or adaptively modifying them based on performance), improving measurement (higher sampling rate, precision, and accuracy, and low-pass filtering if necessary), or cascading multiple PID controllers.

Linearity

Another problem faced with PID controllers is that they are linear, and in particular symmetric. Thus, performance of PID controllers in non-linear systems (such as HVAC systems) is variable. For example, in temperature control, a common use case is active heating (via a heating element) but passive cooling (heating off, but no cooling), so overshoot can only be corrected slowly – it cannot be forced downward. In this case the PID should be tuned to be overdamped, to prevent or reduce overshoot, though this reduces performance (it increases settling time).

Noise in Derivative

A problem with the derivative term is that it amplifies higher frequency measurement or process noise that can cause large amounts of change in the output. It is often helpful to filter the measurements with a low-pass filter in order to remove higher-frequency noise components. As low-pass filtering and derivative control can cancel each other out, the amount of filtering is limited. So low noise instrumentation can be important. A nonlinear median filter may be used, which improves the filtering efficiency and practical performance. In some cases, the differential band can be turned off with little loss of control. This is equivalent to using the PID controller as a PI controller.

Modifications to the PID Algorithm

The basic PID algorithm presents some challenges in control applications that have been addressed by minor modifications to the PID form.

Integral Windup

One common problem resulting from the ideal PID implementations is integral windup. Following a large change in setpoint the integral term can accumulate an error larger than the maximal value for the regulation variable (windup), thus the system overshoots and continues to increase until this accumulated error is unwound. This problem can be addressed by:

- Disabling the integration until the PV has entered the controllable region

- Preventing the integral term from accumulating above or below pre-determined bounds

- Back-calculating the integral term to constrain the regulator output within feasible bounds.

Overshooting from Known Disturbances

For example, a PID loop is used to control the temperature of an electric resistance furnace where the system has stabilized. Now when the door is opened and something cold is put into the furnace the temperature drops below the setpoint. The integral function of the controller tends to compensate for error by introducing another error in the positive direction. This overshoot can be avoided by freezing of the integral function after the opening of the door for the time the control loop typically needs to reheat the furnace.

PI Controller

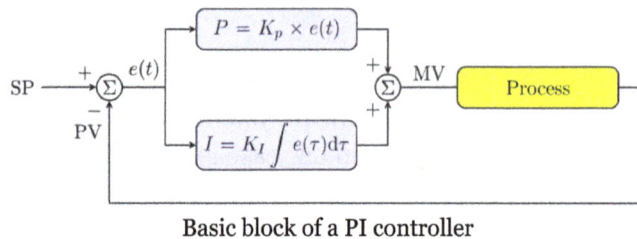

Basic block of a PI controller

A PI Controller (proportional-integral controller) is a special case of the PID controller in which the derivative (D) of the error is not used.

The controller output is given by

$$K_p \Delta + K_I \int \Delta dt$$

where Δ is the error or deviation of actual measured value (PV) from the setpoint (SP).

$$\Delta = SP - PV.$$

A PI controller can be modelled easily in software such as Simulink or Xcos using a "flow chart" box involving Laplace operators:

$$C = \frac{G(1 + \tau s)}{\tau s}$$

where

$G = K_p =$ proportional gain

$G / \tau = K_I =$ integral gain

Setting a value for G is often a trade off between decreasing overshoot and increasing settling time.

The lack of derivative action may make the system more steady in the steady state in the case of noisy data. This is because derivative action is more sensitive to higher-frequency terms in the inputs.

Without derivative action, a PI-controlled system is less responsive to real (non-noise) and relatively fast alterations in state and so the system will be slower to reach setpoint and slower to respond to perturbations than a well-tuned PID system may be.

Deadband

Many PID loops control a mechanical device (for example, a valve). Mechanical maintenance can be a major cost and wear leads to control degradation in the form of either stiction or a deadband in the mechanical response to an input signal. The rate of mechanical wear is mainly a function of how often a device is activated to make a change. Where wear is a significant concern, the PID loop may have an output deadband to reduce the frequency of activation of the output (valve). This is accomplished by modifying the controller to hold its output steady if the change would be small (within the defined deadband range). The calculated output must leave the deadband before the actual output will change.

Setpoint Step Change

The proportional and derivative terms can produce excessive movement in the output when a system is subjected to an instantaneous step increase in the error, such as a large setpoint change. In the case of the derivative term, this is due to taking the derivative of the error, which is very large in the case of an instantaneous step change. As a result, some PID algorithms incorporate some of the following modifications:

Setpoint ramping

> In this modification, the setpoint is gradually moved from its old value to a newly specified value using a linear or first order differential ramp function. This avoids the discontinuity present in a simple step change.

Derivative of the process variable

> In this case the PID controller measures the derivative of the measured process variable (PV), rather than the derivative of the error. This quantity is always continuous (i.e., never has a step change as a result of changed setpoint). This modification is a simple case of setpoint weighting.

Setpoint weighting

> Setpoint weighting adds adjustable factors (usually between 0 and 1) to the setpoint in the error in the proportional and derivative element of the controller. The error in the integral term must be the true control error to avoid steady-state control errors. These two extra parameters do not affect the response to load disturbances and measurement noise and can be tuned to improve the controller's setpoint response.

Feed-forward

The control system performance can be improved by combining the feedback (or closed-loop) control of a PID controller with feed-forward (or open-loop) control. Knowledge about the system (such as the desired acceleration and inertia) can be fed forward and combined with the PID output to improve the overall system performance. The feed-forward value alone can often provide the major portion of the controller output. The PID controller primarily has to compensate whatever difference or *error* remains between the setpoint (SP) and the system response to the open loop control. Since the feed-forward output is not affected by the process feedback, it can never cause the control system to oscillate, thus improving the system response without affecting stability. Feed forward can be based on the setpoint and on extra measured disturbances. Setpoint weighting is a simple form of feed forward.

For example, in most motion control systems, in order to accelerate a mechanical load under control, more force is required from the actuator. If a velocity loop PID controller is being used to control the speed of the load and command the force being applied by the actuator, then it is beneficial to take the desired instantaneous acceleration, scale that value appropriately and add it to the output of the PID velocity loop controller. This means that whenever the load is being accelerated or decelerated, a proportional amount of force is commanded from the actuator regardless of the feedback value. The PID loop in this situation uses the feedback information to change the combined output to reduce the remaining difference between the process setpoint and the feedback value. Working together, the combined open-loop feed-forward controller and closed-loop PID controller can provide a more responsive control system.

Bumpless Operation

PID controllers are often implemented with a "bumpless" initialization feature that recalculates the integral accumulator term to maintain a consistent process output through parameter changes. A partial implementation is to store the integral of the integral gain times the error rather than storing the integral of the error and postmultiplying by the integral gain, which prevents discontinuous output when the I gain is changed, but not the P or D gains.

Other Improvements

In addition to feed-forward, PID controllers are often enhanced through methods such as PID gain scheduling (changing parameters in different operating conditions), fuzzy logic or computational verb logic. Further practical application issues can arise from instrumentation connected to the controller. A high enough sampling rate, measurement precision, and measurement accuracy are required to achieve adequate control performance. Another new method for improvement of PID controller is to increase the degree of freedom by using fractional order. The order of the integrator and differentiator add increased flexibility to the controller.

Cascade Control

One distinctive advantage of PID controllers is that two PID controllers can be used together to yield better dynamic performance. This is called cascaded PID control. In cascade control there are two PIDs arranged with one PID controlling the setpoint of another. A PID controller acts as outer loop controller, which controls the primary physical parameter, such as fluid level or velocity. The other controller acts as inner loop controller, which reads the output of outer loop controller as setpoint, usually controlling a more rapid changing parameter, flowrate or acceleration. It can be mathematically proven that the working frequency of the controller is increased and the time constant of the object is reduced by using cascaded PID controllers.

For example, a temperature-controlled circulating bath has two PID controllers in cascade, each with its own thermocouple temperature sensor. The outer controller controls the temperature of the water using a thermocouple located far from the heater where it accurately reads the temperature of the bulk of the water. The error term of this PID controller is the difference between the desired bath temperature and measured temperature. Instead of controlling the heater directly, the outer PID controller sets a heater temperature goal for the inner PID controller. The inner PID controller controls the temperature of the heater using a thermocouple attached to the heater. The inner controller's error term is the difference between this heater temperature setpoint and the measured temperature of the heater. Its output controls the actual heater to stay near this setpoint.

The proportional, integral and differential terms of the two controllers will be very different. The outer PID controller has a long time constant – all the water in the tank needs to heat up or cool down. The inner loop responds much more quickly. Each controller can be tuned to match the physics of the system *it* controls – heat transfer and thermal mass of the whole tank or of just the heater – giving better total response.

Alternative Nomenclature and PID Forms

Ideal Versus Standard PID Form

The form of the PID controller most often encountered in industry, and the one most relevant to tuning algorithms is the *standard form*. In this form the K_p gain is applied to the I_{out}, and D_{out} terms, yielding:

$$\text{MV(t)} = K_p \left(e(t) + \frac{1}{T_i} \int_0^t e(\tau) d\tau + T_d \frac{d}{dt} e(t) \right)$$

where

   T_i is the *integral time*

   T_d is the *derivative time*

In this standard form, the parameters have a clear physical meaning. In particular, the inner summation produces a new single error value which is compensated for future and past errors. The addition of the proportional and derivative components effectively predicts the error value at T_d seconds (or samples) in the future, assuming that the loop control remains unchanged. The integral component adjusts the error value to compensate for the sum of all past errors, with the intention of completely eliminating them in T_i seconds (or samples). The resulting compensated single error value is scaled by the single gain K_p.

In the ideal parallel form, shown in the controller theory section

$$U(t) = K_p e(t) + K_i \int_0^t e(\tau) d\tau + K_d \frac{d}{dt} e(t)$$

the gain parameters are related to the parameters of the standard form through

$$K_i = \frac{K_p}{T_i} \text{ and } K_d = K_p T_d \; .$$

This parallel form, where the parameters are treated as simple gains, is the most general and flexible form. However, it is also the form where the parameters have the least physical interpretation and is generally reserved for theoretical treatment of the PID controller. The standard form, despite being slightly more complex mathematically, is more common in industry.

Reciprocal Gain

In many cases, the manipulated variable output by the PID controller is a dimensionless fraction between 0 and 100% of some maximum possible value, and the translation into real units (such as pumping rate or watts of heater power) is outside the PID controller. The process variable, however, is in dimensioned units such as temperature. It is common in this case to express the gain K_p not as "output per degree", but rather in the form of a temperature $1/K_p$ which is "degrees per full output". This is the range over which the output changes from 0 to 1 (0% to 100%).

Basing Derivative Action on PV

In most commercial control systems, derivative action is based on PV rather than error. This is because the digitized version of the algorithm produces a large unwanted spike when the SP is changed. If the SP is constant then changes in PV will be the same as changes in error. Therefore, this modification makes no difference to the way the controller responds to process disturbances.

$$MV(t) = K_p\left(e(t) + \frac{1}{T_i}\int_0^t e(\tau)d\tau - T_d \frac{d}{dt}PV(t) \right)$$

Basing Proportional Action on PV

Most commercial control systems offer the option of also basing the proportional action on PV. This means that only the integral action responds to changes in SP. The modification to the algorithm does not affect the way the controller responds to process disturbances. The change to proportional action on PV eliminates the instant and possibly very large change in output on a fast change in SP. Depending on the process and tuning this may be beneficial to the response to a SP step.

$$MV(t) = K_p\left(-PV(t) + \frac{1}{T_i}\int_0^t e(\tau)d\tau - T_d \frac{d}{dt}PV(t) \right)$$

King describes an effective chart-based method.

Laplace form of the PID Controller

Sometimes it is useful to write the PID regulator in Laplace transform form:

$$G(s) = K_p + \frac{K_i}{s} + K_d s = \frac{K_d s^2 + K_p s + K_i}{s}$$

Having the PID controller written in Laplace form and having the transfer function of the controlled system makes it easy to determine the closed-loop transfer function of the system.

PID Pole Zero Cancellation

The PID equation can be written in this form:

$$G(s) = K_d \frac{s^2 + \frac{K_p}{K_d}s + \frac{K_i}{K_d}}{s}$$

When this form is used it is easy to determine the closed loop transfer function.

$$H(s) = \frac{1}{s^2 + 2\zeta\omega_0 s + \omega_0^2}$$

If

$$\frac{K_i}{K_d} = \omega_0^2$$

$$\frac{K_p}{K_d} = 2\zeta\omega_0$$

Then

$$G(s)H(s) = \frac{K_d}{s}$$

While this appears to be very useful to remove unstable poles, it is in reality not the case. The closed loop transfer function from disturbance to output still contains the unstable poles.

Series/interacting form

Another representation of the PID controller is the series, or *interacting* form

$$G(s) = K_c \frac{(\tau_i s + 1)}{\tau_i s}(\tau_d s + 1)$$

where the parameters are related to the parameters of the standard form through

$$K_p = K_c \cdot \alpha, \; T_i = \tau_i \cdot \alpha, \text{ and}$$

$$T_d = \frac{\tau_d}{\alpha}$$

with

$$\alpha = 1 + \frac{\tau_d}{\tau_i}.$$

This form essentially consists of a PD and PI controller in series, and it made early (analog) controllers easier to build. When the controllers later became digital, many kept using the interacting form.

Discrete Implementation

The analysis for designing a digital implementation of a PID controller in a microcontroller (MCU) or FPGA device requires the standard form of the PID controller to be *discretized*. Approximations for first-order derivatives are made by backward finite differences. The integral term is discretised, with a sampling time Δt, as follows,

$$\int_0^{t_k} e(\tau)d\tau = \sum_{i=1}^{k} e(t_i)\Delta t$$

The derivative term is approximated as,

$$\frac{de(t_k)}{dt} = \frac{e(t_k) - e(t_{k-1})}{\Delta t}$$

Thus, a *velocity algorithm* for implementation of the discretized PID controller in a MCU is obtained by differentiating $u(t)$, using the numerical definitions of the first and second derivative and solving for $u(t_k)$ and finally obtaining:

$$u(t_k) = u(t_{k-1}) + K_p \left[\left(1 + \frac{\Delta t}{T_i} + \frac{T_d}{\Delta t} \right) e(t_k) + \left(-1 - \frac{2T_d}{\Delta t} \right) e(t_{k-1}) + \frac{T_d}{\Delta t} e(t_{k-2}) \right]$$

$$\text{s.t. } T_i = K_p / K_i, T_d = K_d / K_p$$

Pseudocode

Here is a simple software loop that implements a PID algorithm:

previous_error = 0

integral = 0

start:

error = setpoint - measured_value

integral = integral + error*dt

derivative = (error - previous_error)/dt

output = Kp*error + Ki*integral + Kd*derivative

previous_error = error

wait(dt)

goto start

In this example, two variables that will be maintained within the loop are initialized to zero, then the loop begins. The current *error* is calculated by subtracting the *measured_value* (the process variable or PV) from the current *setpoint* (SP). Then, *integral* and *derivative* values are calculated and these and the *error* are combined with three preset gain terms – the proportional gain, the integral gain and the derivative gain – to derive an *output* value. In the real world, this is D to A converted and passed into the process under control as the manipulated variable (or MV). The current error is stored elsewhere for re-use in the next differentiation, the program then waits until dt seconds have passed since start, and the loop begins again, reading in new values for the PV and the setpoint and calculating a new value for the error.

Fly-by-wire

Fly-by-wire (FBW) is a system that replaces the conventional manual flight controls of an aircraft with an electronic interface. The movements of flight controls are converted to electronic signals transmitted by wires (hence the fly-by-wire term), and flight control computers determine how to move the actuators at each control surface to provide the ordered response. The fly-by-wire system also allows automatic signals sent by the aircraft's computers to perform functions without the pilot's input, as in systems that automatically help stabilize the aircraft, or prevent unsafe operation of the aircraft outside of its performance envelope.

Development

Mechanical and hydro-mechanical flight control systems are relatively heavy and require careful routing of flight control cables through the aircraft by systems of pulleys, cranks, tension cables and hydraulic pipes. Both systems often require redundant backup to deal with failures, which increases weight. Both have limited ability to compensate for changing aerodynamic conditions. Dangerous characteristics such as stalling, spinning and pilot-induced oscillation (PIO), which depend mainly on the stability and structure of the aircraft concerned rather than the control system itself, can still occur with these systems.

The term "fly-by-wire" implies a purely electrically signaled control system. It is used in the general sense of computer-configured controls, where a computer system is interposed between the operator and the final control actuators or surfaces. This modifies the manual inputs of the pilot in accordance with control parameters.

Side-sticks, centre sticks, or conventional flight control yokes can be used to fly FBW aircraft.

Basic Operation

Command

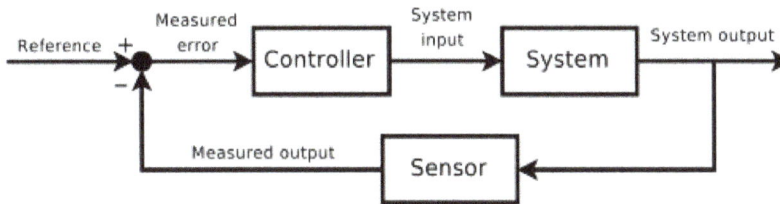

Simple feedback loop

Fly-by wire systems are quite complex, but their operation can be explained in simple terms. When a pilot moves the control column (or sidestick), a signal is sent to a computer (analogous to moving a game controller) the signal is sent through multiple wires (channels) to ensure that the signal reaches the computer. A 'Triplex' is when there are three channels being used. In an Analog system, the computer receives the signals, performs a calculation (adds the signal voltages and divides by the number of signals received to find the mean average voltage) and adds another channel. These four 'Quadruplex' signals are then sent to the control surface actuator, and the surface begins to move. Potentiometers in the actuator send a signal back to the computer (usually a negative voltage) reporting the position of the actuator. When the actuator reaches the desired position, the two signals (incoming and outgoing) cancel each other out and the actuator stops moving (completing a feedback loop). In a Digital Fly By Wire Flight Control System complex software interprets digital signals from the pilots control input sensors and performs calculations based on the Flight Control Laws programmed into the Flight Control Computers and input from the Air Data Inertial Reference Units and other sensors. The computer then commands the flight control surfaces to adopt a configuration that will achieve the desired flight path.

Automatic Stability Systems

Fly-by-wire control systems allow aircraft computers to perform tasks without pilot input. Automatic stability systems operate in this way. Gyroscopes fitted with sensors are mounted in an aircraft to sense movement changes in the pitch, roll and yaw axes. Any movement (from straight and level flight for example) results in signals to the computer, which automatically moves control actuators to stabilize the aircraft.

Safety and Redundancy

Aircraft systems may be quadruplexed (four independent channels) to prevent loss of signals in the case of failure of one or even two channels. High performance aircraft that have fly-by-wire controls (also called CCVs or Control-Configured Vehicles) may be deliberately designed to have low or even negative stability in some flight regimes—the rapid-reacting CCV controls compensating for the lack of natural stability.

Pre-flight safety checks of a fly-by-wire system are often performed using built-in test equipment (BITE). On programming the system, either by the pilot or groundcrew, a number of control movement steps are automatically performed. Any failure will be indicated to the crews.

Some aircraft, the Panavia Tornado for example, retain a very basic hydro-mechanical backup system for limited flight control capability on losing electrical power; in the case of the Tornado this allows rudimentary control of the stabilators only for pitch and roll axis movements.

Weight Saving

A FBW aircraft can be lighter than a similar design with conventional controls. This is partly due to the lower overall weight of the system components, and partly because the natural stability of the aircraft can be relaxed, slightly for a transport aircraft and more for a maneuverable fighter, which means that the stability surfaces that are part of the aircraft structure can therefore be made smaller. These include the vertical and horizontal stabilizers (fin and tailplane) that are (normally) at the rear of the fuselage. If these structures can be reduced in size, airframe weight is reduced. The advantages of FBW controls were first exploited by the military and then in the commercial airline market. The Airbus series of airliners used full-authority FBW controls beginning with their A320 series, see A320 flight control (though some limited FBW functions existed on A310). Boeing followed with their 777 and later designs.

Electronic fly-by-wire systems can respond flexibly to changing aerodynamic conditions, by tailoring flight control surface movements so that aircraft response to control inputs is appropriate to flight conditions. Electronic systems require less maintenance, whereas mechanical and hydraulic systems require lubrication, tension adjustments, leak checks, fluid changes, etc. Placing circuitry between pilot and aircraft can enhance safety. For example, the control system can try to prevent a stall, or it can stop the pilot from over stressing the airframe.

The main concern with fly-by-wire systems is reliability. While traditional mechanical or hydraulic control systems usually fail gradually, the loss of all flight control computers could immediately render the aircraft uncontrollable. For this reason, most fly-by-wire systems incorporate either redundant computers (triplex, quadruplex etc.), some kind of mechanical or hydraulic backup or a combination of both. A "mixed" control system such as the latter is not desirable and modern FBW aircraft normally avoid it by having more independent FBW channels, thereby reducing the possibility of overall failure to minuscule levels that are acceptable to the independent regulatory and safety authority responsible for aircraft design, testing and certification before operational service.

History

Avro Canada CF-105 Arrow, first non-experimental aircraft flown with a fly-by-wire control system

F-8C Crusader digital fly-by-wire testbed

Servo-electrically operated control surfaces was first tested in the 1930s on the Soviet Tupolev ANT-20. Long runs of mechanical and hydraulic connections were replaced with wires and electric servos.

The first pure electronic fly-by-wire aircraft with no mechanical or hydraulic backup was the Apollo Lunar Landing Research Vehicle (LLRV), first flown in 1964.

The first non-experimental aircraft that was designed and flown (in 1958) with a fly-by-wire flight control system was the Avro Canada CF-105 Arrow, a feat not repeated with a production aircraft until Concorde in 1969. This system also included solid-state components and system redundancy, was designed to be integrated with a computerised navigation and automatic search and track radar, was flyable from ground control with data uplink and downlink, and provided artificial feel (feedback) to the pilot.

In the UK the two seater Avro 707B was flown with a Fairey system with mechanical backup in the early to mid-60s. The programme was curtailed when the airframe ran out of flight time.

The first digital fly-by-wire fixed-wing aircraft without a mechanical backup to take to the air (in 1972) was an F-8 Crusader, which had been modified electronically by NASA of the United States as a test aircraft. This was preceded in 1964 by the LLRV which pioneered fly-by-wire flight with no mechanical backup. Control was through a digital computer with three analogue redundant channels. In the USSR the Sukhoi T-4 also flew. At about the same time in the United Kingdom a trainer variant of the British Hawker Hunter fighter was modified at the British Royal Aircraft Establishment with fly-by-wire flight controls for the right-seat pilot. This was test-flown, with the left-seat pilot having conventional flight controls for safety reasons, and with the capability for him to override and turn off the fly-by-wire system. It flew in April 1972.

Analog Systems

All "fly-by-wire" flight control systems eliminate the complexity, the fragility, and the weight of the mechanical circuit of the hydromechanical or electromechanical flight control systems— each being replaced with electronic circuits. The control mechanisms in the cockpit now operate signal transducers, which in turn generate the appropriate electronic commands. These are next processed by an electronic controller—either an analog one, or (more modernly) a digital one. Aircraft and spacecraft autopilots are now part of the electronic controller.

The hydraulic circuits are similar except that mechanical servo valves are replaced with electrically controlled servo valves, operated by the electronic controller. This is the simplest and earliest configuration of an analog fly-by-wire flight control system. In this configuration, the flight control systems must simulate "feel". The electronic controller controls electrical feel devices that provide the appropriate "feel" forces on the manual controls. This was used in Concorde, the first production fly-by-wire airliner.

In more sophisticated versions, analog computers replaced the electronic controller. The canceled 1950s Canadian supersonic interceptor, the Avro Canada CF-105 Arrow, employed this type of system. Analog computers also allowed some customization of flight control characteristics, including relaxed stability. This was exploited by the early versions of F-16, giving it impressive maneuverability.

Digital Systems

The Airbus A320, first airliner with digital fly-by-wire controls

A digital fly-by-wire flight control system is similar to its analog counterpart. However, the signal processing is done by digital computers and the pilot literally can "fly-via-computer". This also increases the flexibility of the flight control system, since the digital computers can receive input from any aircraft sensor (such as the altimeters and the pitot tubes). This also increases the electronic stability, because the system is less dependent on the values of critical electrical components in an analog controller.

The computers sense position and force inputs from pilot controls and aircraft sensors. They solve differential equations to determine the appropriate command signals that move the flight controls to execute the intentions of the pilot.

The programming of the digital computers enable flight envelope protection. These protections are tailored to an aircraft's handling characteristics to stay within aerodynamic and structural limitations of the aircraft. For example, the computer in flight envelope protection mode can try to prevent the aircraft from being handled dangerously by preventing pilots from exceeding preset limits on the aircraft's flight-control envelope, such as those that prevent stalls and spins, and which limit airspeeds and g forces on the airplane. Software can also be included that stabilize the flight-control inputs to avoid pilot-induced oscillations.

Since the flight-control computers continuously "fly" the aircraft, pilot's workloads can be reduced. Also, in military and naval applications, it is now possible to fly military aircraft that have relaxed stability. The primary benefit for such aircraft is more maneuverability during combat and training flights, and the so-called "carefree handling" because stalling, spinning and other undesirable performances are prevented automatically by the computers.

Digital flight control systems enable inherently unstable combat aircraft, such as the Lockheed F-117 Nighthawk and the Northrop Grumman B-2 Spirit flying wing to fly in usable and safe manners.

Applications

A Dassault Falcon 7X, the first business jet with digital fly-by-wire controls

- The Space Shuttle orbiter has an all-digital fly-by-wire control system. This system was first exercised (as the only flight control system) during the glider unpowered-flight "Approach and Landing Tests" that began on the Space Shuttle *Enterprise* during 1977.

- Launched into production during 1984, the Airbus Industries Airbus A320 became the first airliner to fly with an all-digital fly-by-wire control system.

- During 2005, the Dassault Falcon 7X became the first business jet with fly-by-wire controls.

Legislation

The Federal Aviation Administration (FAA) of the United States has adopted the RTCA/DO-

178B, titled "Software Considerations in Airborne Systems and Equipment Certification", as the certification standard for aviation software. Any safety-critical component in a digital fly-by-wire system including applications of the laws of aeronautics and computer operating systems will need to be certified to DO-178B Level A, which is applicable for preventing potential catastrophic failures.

Nevertheless, the top concern for computerized, digital, fly-by-wire systems is reliability, even more so than for analog electronic control systems. This is because the digital computers that are running software are often the only control path between the pilot and aircraft's flight control surfaces. If the computer software crashes for any reason, the pilot may be unable to control an aircraft. Hence virtually all fly-by-wire flight control systems are either triply or quadruply redundant in their computers and electronics. These have three or four flight-control computers operating in parallel, and three or four separate data buses connecting them with each control surface.

Redundancy

If one of the flight-control computers crashes, or is damaged in combat, or suffers from "insanity" caused by electromagnetic pulses, the others overrule the faulty one (or even two of them), they continue flying the aircraft safely, and they can either turn off or re-boot the faulty computers. Any flight-control computer whose results disagree with the others is ruled to be faulty, and it is either ignored or re-booted. (In other words, it is voted-out of control by the others.)

In addition, most of the early digital fly-by-wire aircraft also had an analog electrical, a mechanical, or a hydraulic back-up flight control system. The Space Shuttle has, in addition to its redundant set of four digital computers running its primary flight-control software, a fifth back-up computer running a separately developed, reduced-function, software flight-control system – one that can be commanded to take over in the event that a fault ever affects all of the computers in the other four. This back-up system serves to reduce the risk of total flight-control-system failure ever happening because of a general-purpose flight software fault that has escaped notice in the other four computers.

For airliners, flight-control redundancy improves their safety, but fly-by-wire control systems also improve economy in flight because they are lighter, and they eliminate the need for many mechanical, and heavy, flight-control mechanisms. Furthermore, most modern airliners have computerized systems that control their jet engine throttles, air inlets, fuel storage and distribution system, in such a way to minimize their consumption of jet fuel. Thus, digital control systems do their best to reduce the cost of flights.

Airbus/Boeing

Airbus and Boeing commercial airplanes differ in their approaches in using fly-by-wire systems. In Airbus airliners, the flight-envelope control system always retains ultimate flight control when flying under normal law, and it will not permit the pilots to fly outside these performance limits unless flying under alternate law. However, in the event of multiple failures of redundant computers, the A320 does have a mechanical back-up system for its pitch trim and its rudder. The A340-600 has a purely electrical (not electronic) back-up rudder control system, and beginning with the new A380 airliner, all flight-control systems have back-up systems that are purely electrical through the use of a so-called "three-axis Backup Control Module" (BCM)

With the Boeing 777 model airliners, the two pilots can completely override the computerized flight-control system to permit the aircraft to be flown beyond its usual flight-control envelope during emergencies. Airbus's strategy, which began with the Airbus A320, has been continued on subsequent Airbus airliners.

Engine Digital Control

The advent of FADEC (Full Authority Digital Engine Control) engines permits operation of the flight control systems and autothrottles for the engines to be fully integrated. On modern military aircraft other systems such as autostabilization, navigation, radar and weapons system are all integrated with the flight control systems. FADEC allows maximum performance to be extracted from the aircraft without fear of engine misoperation, aircraft damage or high pilot workloads.

In the civil field, the integration increases flight safety and economy. The Airbus A320 and its fly-by-wire brethren are protected from dangerous situations such as low-speed stall or overstressing by flight envelope protection. As a result, in such conditions, the flight control systems commands the engines to increase thrust without pilot intervention. In economy cruise modes, the flight control systems adjust the throttles and fuel tank selections more precisely than all but the most skillful pilots. FADEC reduces rudder drag needed to compensate for sideways flight from unbalanced engine thrust. On the A330/A340 family, fuel is transferred between the main (wing and center fuselage) tanks and a fuel tank in the horizontal stabilizer, to optimize the aircraft's center of gravity during cruise flight. The fuel management controls keep the aircraft's center of gravity accurately trimmed with fuel weight, rather than drag-inducing aerodynamic trims in the elevators.

Further Developments

Fly-by-optics

Fly-by-optics is sometimes used instead of fly-by-wire because it offers a higher data transfer rate, immunity to electromagnetic interference, and lighter weight. In most cases, the cables are just changed from electrical to optical fiber cables. Sometimes it is referred to as "fly-by-light" due to its use of fiber optics. The data generated by the software and interpreted by the controller remain the same.

Power-by-wire

Having eliminated the mechanical transmission circuits in fly-by-wire flight control systems, the next step is to eliminate the bulky and heavy hydraulic circuits. The hydraulic circuit is replaced by an electrical power circuit. The power circuits power electrical or self-contained electrohydraulic actuators that are controlled by the digital flight control computers. All benefits of digital fly-by-wire are retained.

The biggest benefits are weight savings, the possibility of redundant power circuits and tighter integration between the aircraft flight control systems and its avionics systems. The absence of hydraulics greatly reduces maintenance costs. This system is used in the Lockheed Martin F-35

Lightning II and in Airbus A380 backup flight controls. The Boeing 787 will also incorporate some electrically operated flight controls (spoilers and horizontal stabilizer), which will remain operational with either a total hydraulics failure and/or flight control computer failure.

Fly-by-wireless

Wiring adds a considerable amount of weight to an aircraft; therefore, researchers are exploring implementing fly-by-wireless solutions. Fly-by-wireless systems are very similar to fly-by-wire systems, however, instead of using a wired protocol for the physical layer a wireless protocol is employed.

In addition to reducing weight, implementing a wireless solution has the potential to reduce costs throughout an aircraft's life cycle. For example, many key failure points associated with wire and connectors will be eliminated thus hours spent troubleshooting wires and connectors will be reduced. Furthermore, engineering costs could potentially decrease because less time would be spent on designing wiring installations, late changes in an aircraft's design would be easier to manage, etc.

Intelligent Flight Control System

A newer flight control system, called intelligent flight control system (IFCS), is an extension of modern digital fly-by-wire flight control systems. The aim is to intelligently compensate for aircraft damage and failure during flight, such as automatically using engine thrust and other avionics to compensate for severe failures such as loss of hydraulics, loss of rudder, loss of ailerons, loss of an engine, etc. Several demonstrations were made on a flight simulator where a Cessna-trained small-aircraft pilot successfully landed a heavily damaged full-size concept jet, without prior experience with large-body jet aircraft. This development is being spearheaded by NASA Dryden Flight Research Center. It is reported that enhancements are mostly software upgrades to existing fully computerized digital fly-by-wire flight control systems.

Distributed Control System

A distributed control system (DCS) is a control system for a process or plant, wherein control elements are distributed throughout the system. This is in contrast to non-distributed systems, which use a single controller at a central location. In a DCS, a hierarchy of controllers is connected by communications networks for command and monitoring.

Example scenarios where a DCS might be used include:

- Chemical plants

- Petrochemical (oil) and refineries

- Pulp and Paper Mills

- Boiler controls and power plant systems

- Nuclear power plants

- Environmental control systems

- Water management systems

- Water treatment plants

- Sewage treatment plants

- Food and food processing

- Agro chemical and fertilizer

- Metal and mines

- Automobile manufacturing

- Metallurgical process plants

- Pharmaceutical manufacturing

- Sugar refining plants

- Dry cargo and bulk oil carrier ships

- Formation control of multi-agent systems

Elements

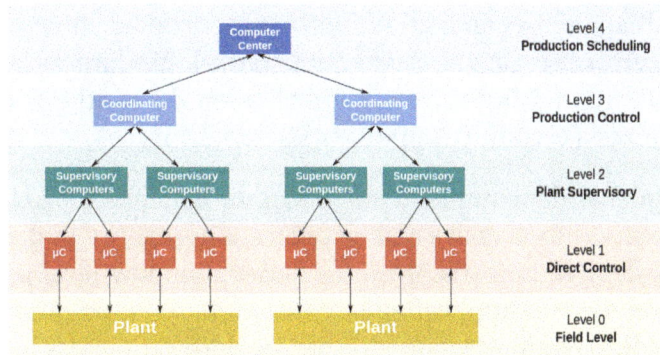

Functional levels of a typical Distributed Control System.

A DCS typically uses custom designed processors as controllers and uses both proprietary interconnections or standard communications protocol for communication. Input and output modules form component parts of the DCS. The processor receives information from input modules, processing the information and decides actions to be performed by the output modules. The input modules receive information from input instruments in the process (or field) mainly via sensors and the output modules transmit instructions to the instruments in the field for initiating actions mainly via final control elements . The inputs and outputs can be either analog signal which are continuously changing or discrete signals which are 2 state either on or off, the signals can also be via fieldbus such as foundation fieldbus, profibus, HART, Modbus and other digital communication buses with communicates not only input and output signals but also advance messages such as error diagnostics and status signals.

The elements of a DCS may connect directly to physical equipment such as switches, pumps and valves and to Human Machine Interface (HMI).

Applications

Distributed control systems (DCSs) are dedicated systems used to control manufacturing processes that are continuous or batch-oriented, such as oil refining, petrochemicals, central station power generation, fertilizers, pharmaceuticals, food and beverage manufacturing, cement production, steelmaking, and papermaking. DCSs are connected to sensors and actuators and use setpoint control to control the flow of material through the plant. The most common example is a setpoint control loop consisting of a pressure sensor, controller, and control valve. Pressure or flow measurements are transmitted to the controller, usually through the aid of a signal conditioning input/output (I/O) device. When the measured variable reaches a certain point, the controller instructs a valve or actuation device to open or close until the fluidic flow process reaches the desired setpoint. Large oil refineries have many thousands of I/O points and employ very large DCSs. Processes are not limited to fluidic flow through pipes, however, and can also include things like paper machines and their associated quality controls, variable speed drives and motor control centers, cement kilns, mining operations, ore processing facilities, and many others.

A typical DCS consists of functionally and/or geographically distributed digital controllers capable of executing from regulatory control loops in one control box. The input/output devices (I/O) can be integral with the controller or located remotely via a field network. Today's controllers have extensive computational capabilities and, in addition to proportional, integral, and derivative (PID) control, can generally perform logic and sequential control. Modern DCSs also support neural networks and fuzzy application. Recent research focuses on the synthesis of optimal distributed controllers, which optimizes a certain H-infinity or H-2 criterion.

DCSs are usually designed with redundant processors to enhance the reliability of the control system. Most systems come with displays and configuration software that enable the end-user to configure the control system without the need for performing low-level programming, allowing the user also to better focus on the application rather than the equipment. However, considerable system knowledge and skill is required to properly deploy the hardware, software, and applications. Many plants have dedicated personnel who focus on these tasks, augmented by vendor support that may include maintenance support contracts.

DCSs may employ one or more workstations and can be configured at the workstation or by an off-line personal computer. Local communication is handled by a control network with transmission over twisted -pair, coaxial, or fiber-optic cable. A server and/or applications processor may be included in the system for extra computational, data collection, and reporting capability.

History

Early minicomputers were used in the control of industrial processes since the beginning of the

1960s. The IBM 1800, for example, was an early computer that had input/output hardware to gather process signals in a plant for conversion from field contact levels (for digital points) and analog signals to the digital domain.

Manual controls in 1958 industry, photo by Paolo Monti

The first industrial control computer system was built 1959 at the Texaco Port Arthur, Texas, refinery with an RW-300 of the Ramo-Wooldridge Company

In 1975, both Honeywell and Japanese electrical engineering firm Yokogawa introduced their own independently produced DCS's with Yokogawa introducing and successfully installing before Honeywell, with the TDC 2000 and CENTUM systems, respectively. US-based Bristol also introduced their UCS 3000 universal controller in 1975. In 1978 Valmet introduced their own DCS system called Damatic (latest generation named Valmet DNA). In 1980, Bailey (now part of ABB) introduced the NETWORK 90 system, Fisher Controls (now part of Emerson Electric) introduced the PROVoX system, Fischer & Porter Company (now also part of ABB) introduced DCI-4000 (DCI stands for Distributed Control Instrumentation).

The DCS largely came about due to the increased availability of microcomputers and the proliferation of microprocessors in the world of process control. Computers had already been applied to process automation for some time in the form of both direct digital control (DDC) and set point control. In the early 1970s Taylor Instrument Company, (now part of ABB) developed the 1010 system, Foxboro the FOX1 system, Fisher Controls the DC^2 system and Bailey Controls the 1055 systems. All of these were DDC applications implemented within minicomputers (DEC PDP-11, Varian Data Machines, MODCOMP etc.) and connected to proprietary Input/Output hardware. Sophisticated (for the time) continuous as well as batch control was implemented in this way. A more conservative approach was set point control, where process computers supervised clusters of analog process controllers. A CRT-based workstation provided visibility into the process using text and crude character graphics. Availability of a fully functional graphical user interface was a way away.

Central to the DCS model was the inclusion of control function blocks. Function blocks evolved from early, more primitive DDC concepts of "Table Driven" software. One of the first embodiments of object-oriented software, function blocks were self-contained "blocks" of code that emulated analog hardware control components and performed tasks that were essential to process control, such as execution of PID algorithms. Function blocks continue to endure as the predominant method of control for DCS suppliers, and are supported by key technologies such as Foundation Fieldbus today.

Midac Systems, of Sydney, Australia, developed an objected-oriented distributed direct digital control system in 1982. The central system ran 11 microprocessors sharing tasks and common memory and connected to a serial communication network of distributed controllers each running two Z80s. The system was installed at the University of Melbourne.

Digital communication between distributed controllers, workstations and other computing elements (peer to peer access) was one of the primary advantages of the DCS. Attention was duly focused on the networks, which provided the all-important lines of communication that, for process applications, had to incorporate specific functions such as determinism and redundancy. As a result, many suppliers embraced the IEEE 802.4 networking standard. This decision set the stage for the wave of migrations necessary when information technology moved into process automation and IEEE 802.3 rather than IEEE 802.4 prevailed as the control LAN.

The Network-centric Era of the 1980s

In the 1980s, users began to look at DCSs as more than just basic process control. A very early example of a Direct Digital Control DCS was completed by the Australian business Midac in 1981–82 using R-Tec Australian designed hardware. The system installed at the University of Melbourne used a serial communications network, connecting campus buildings back to a control room "front end". Each remote unit ran two Z80 microprocessors, while the front end ran eleven Z80s in a parallel processing configuration with paged common memory to share tasks and that could run up to 20,000 concurrent control objects.

It was believed that if openness could be achieved and greater amounts of data could be shared throughout the enterprise that even greater things could be achieved. The first attempts to increase the openness of DCSs resulted in the adoption of the predominant operating system of the day: *UNIX*. UNIX and its companion networking technology TCP-IP were developed by the US Department of Defense for openness, which was precisely the issue the process industries were looking to resolve.

As a result, suppliers also began to adopt Ethernet-based networks with their own proprietary protocol layers. The full TCP/IP standard was not implemented, but the use of Ethernet made it possible to implement the first instances of object management and global data access technology. The 1980s also witnessed the first PLCs integrated into the DCS infrastructure. Plant-wide historians also emerged to capitalize on the extended reach of automation systems. The first DCS supplier to adopt UNIX and Ethernet networking technologies was Foxboro, who introduced the I/A Series system in 1987.

The Application-centric Era of the 1990s

The drive toward openness in the 1980s gained momentum through the 1990s with the increased adoption of commercial off-the-shelf (COTS) components and IT standards. Probably the biggest transition undertaken during this time was the move from the UNIX operating system to the Windows environment. While the realm of the real time operating system (RTOS) for control applications remains dominated by real time commercial variants of UNIX or proprietary operating systems, everything above real-time control has made the transition to Windows.

The introduction of Microsoft at the desktop and server layers resulted in the development of technologies such as OLE for process control (OPC), which is now a de facto industry connectivity standard. Internet technology also began to make its mark in automation and the DCS world, with most DCS HMI supporting Internet connectivity. The 1990s were also known for the "Fieldbus Wars", where rival organizations competed to define what would become the IEC fieldbus standard for digital communication with field instrumentation instead of 4–20 milliamp analog communications. The first fieldbus installations occurred in the 1990s. Towards the end of the decade, the technology began to develop significant momentum, with the market consolidated around Ethernet I/P, Foundation Fieldbus and Profibus PA for process automation applications. Some suppliers built new systems from the ground up to maximize functionality with fieldbus, such as Rockwell PlantPAX System, Honeywell with Experion & Plantscape SCADA systems, ABB with System 800xA, Emerson Process Management with the Emerson Process Management DeltaV control system, Siemens with the SPPA-T3000 or Simatic PCS 7, Forbes Marshall with the Microcon+ control system and Azbil Corporation with the Harmonas-DEO system. Fieldbus technics have been used to integrate machine, drives, quality and condition monitoring applications to one DCS with Valmet DNA system.

The impact of COTS, however, was most pronounced at the hardware layer. For years, the primary business of DCS suppliers had been the supply of large amounts of hardware, particularly I/O and controllers. The initial proliferation of DCSs required the installation of prodigious amounts of this hardware, most of it manufactured from the bottom up by DCS suppliers. Standard computer components from manufacturers such as Intel and Motorola, however, made it cost prohibitive for DCS suppliers to continue making their own components, workstations, and networking hardware.

As the suppliers made the transition to COTS components, they also discovered that the hardware market was shrinking fast. COTS not only resulted in lower manufacturing costs for the supplier, but also steadily decreasing prices for the end users, who were also becoming increasingly vocal over what they perceived to be unduly high hardware costs. Some suppliers that were previously stronger in the PLC business, such as Rockwell Automation and Siemens, were able to leverage their expertise in manufacturing control hardware to enter the DCS marketplace with cost effective offerings, while the stability/scalability/reliability and functionality of these emerging systems are still improving. The traditional DCS suppliers introduced new generation DCS System based on the latest Communication and IEC Standards, which resulting in a trend of combining the traditional concepts/functionalities for PLC and DCS into a one for all solution—named "Process Automation System". The gaps among the various systems remain at the areas such as: the database integrity, pre-engineering functionality, system maturity, communication transparency and reliability. While it is expected the cost ratio is relatively the same (the more powerful the systems are, the more expensive they will be), the reality of the automation business is often operating strategically case by case. The current next evolution step is called Collaborative Process Automation Systems.

To compound the issue, suppliers were also realizing that the hardware market was becoming saturated. The life cycle of hardware components such as I/O and wiring is also typically in the range of 15 to over 20 years, making for a challenging replacement market. Many of the older systems that were installed in the 1970s and 1980s are still in use today, and there is a considerable installed base of systems in the market that are approaching the end of their useful life. Developed industrial economies in North America, Europe, and Japan already had many thousands of DCSs installed, and with few if any new plants being built, the market for new hardware was shifting rapidly to smaller, albeit faster growing regions such as China, Latin America, and Eastern Europe.

Because of the shrinking hardware business, suppliers began to make the challenging transition from a hardware-based business model to one based on software and value-added services. It is a transition that is still being made today. The applications portfolio offered by suppliers expanded considerably in the '90s to include areas such as production management, model-based control, real-time optimization, plant asset management (PAM), Real-time performance management (RPM) tools, alarm management, and many others. To obtain the true value from these applications, however, often requires a considerable service content, which the suppliers also provide.

Modern Systems (2010 onwards)

The latest developments in DCS include the following new technologies:

1. Wireless systems and protocols

2. Remote transmission, logging and data historian

3. Mobile interfaces and controls

4. Embedded web-servers

Increasingly, and ironically, DCS are becoming centralised at plant level, with the ability to log in to the remote equipment. This enables operator to control both at enterprise level (macro) and at the equipment level (micro) both within and outside the plant as physical location due to interconnectivity primarily due to wireless and remote access has shrunk.

As wireless protocols are developed and refined, DCS increasingly includes wireless communication. DCS controllers are now often equipped with embedded servers and provide on-the-go web access. Whether DCS will lead IIOT or borrow key elements from remains to be established.

Many vendors provide the option of a mobile HMI, ready for both Android and iOS. With these interfaces, the threat of security breaches and possible damage to plant and process are now very real.

Networked Control System

A Networked Control System (NCS) is a control system wherein the control loops are closed through a communication network. The defining feature of an NCS is that control and feedback signals are exchanged among the system's components in the form of information packages through a network.

Overview

The functionality of a typical NCS is established by the use of four basic elements:

1. Sensors, to acquire information,

2. Controllers, to provide decision and commands,

3. Actuators, to perform the control commands and

4. Communication network, to enable exchange of information.

The most important feature of a NCS is that it connects cyberspace to physical space enabling the execution of several tasks from long distance. In addition, networked control systems eliminate unnecessary wiring reducing the complexity and the overall cost in designing and implementing the control systems. They can also be easily modified or upgraded by adding sensors, actuators and controllers to them with relatively low cost and no major changes in their structure. Moreover, featuring efficient sharing of data between their controllers, NCS are able to easily fuse global information to make intelligent decisions over large physical spaces.

Their potential applications are numerous and cover a wide range of industries such as: space and terrestrial exploration, access in hazardous environments, factory automation, remote diagnostics and troubleshooting, experimental facilities, domestic robots, aircraft, automobiles, manufacturing plant monitoring, nursing homes and tele-operations. While the potential applications of NCS are numerous, the proven applications are few, and the real opportunity in the area of NCS is in developing real-world applications that realize the area's potential.

Types of Communication Networks

- Fieldbuses, e.g. CAN, LON etc.

- Ethernet

- Wireless networks, e.g. Bluetooth or ZigBee. The term Wireless Networked Control System (WNCS) is often used in this connection.

Problems and Solutions

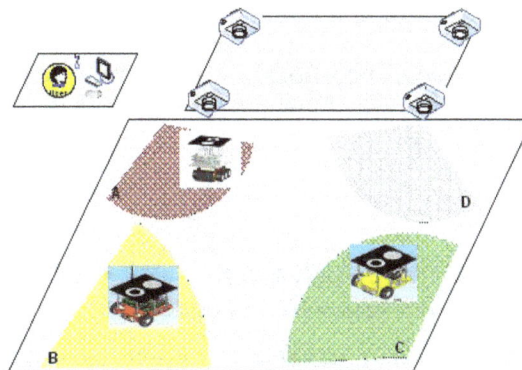

iSpace concept

Advent and development of the Internet combined with the advantages provided by NCS attracted the interest of researchers around the globe. Along with the advantages, several challenges also emerged giving rise to many important research topics. New control strategies, kinematics of the actuators in the systems, reliability and security of communications, bandwidth allocation, development of data communication protocols, corresponding fault detection and fault tolerant control strategies, real-time information collection and efficient processing of sensors data are some of the relative topics studied in depth.

The insertion of the communication network in the feedback control loop makes the analysis and design of an NCS complex, since it imposes additional time delays in control loops or possibility of packages loss. Depending on the application, time-delays could impose severe degratation on the system performance.

To alleviate the time-delay effect, Y. Tipsuwan and M-Y. Chow, in ADAC Lab at North Carolina State University, proposed the Gain Scheduler Middleware (GSM) methodology and applied it in iSpace. S. Munir and W.J. Book (Georgia Institute of Technology) used a Smith predictor, a Kalman filter and an energy regulator to perform teleoperation through the Internet.

K.C. Lee, S. Lee and H.H. Lee used a genetic algorithm to design a controller used in a NCS. Many other researchers provided solutions using concepts from several control areas such as robust control, optimal stochastic control, model predictive control, fuzzy logic etc.

Moreover, a most critical and important issue surrounding the design of distributed NCSs with the successively increasing complexity is to meet the requirements on system reliability and dependability, while guaranteeing a high system performance over a wide operating range. This makes network based fault detection and diagnosis techniques, which are essential to monitor the system performance, receive more and more attention.

Sampled Data System

In systems science, a **sampled-data system** is a control system in which a continuous-time plant is controlled with a digital device. Under periodic sampling, the sampled-data system is time-varying but also periodic; thus, it may be modeled by a simplified discrete-time system obtained by discretizing the plant. However, this discrete model does not capture the inter-sample behavior of the real system, which may be critical in a number of applications.

The analysis of sampled-data systems incorporating full-time information leads to challenging control problems with a rich mathematical structure. Many of these problems have only been solved recently.

Building Management System

A building management system (BMS), otherwise known as a building automation system (BAS), is a computer-based control system installed in buildings that controls and monitors the building's mechanical and electrical equipment such as ventilation, lighting, power systems, fire systems, and security systems. A BMS consists of software and hardware; the software program, usually configured in a hierarchical manner, can be proprietary, using such protocols as C-Bus, Profibus, and so on. Vendors are also producing BMSs that integrate using Internet protocols and open standards such as DeviceNet, SOAP, XML, BACnet, LonWorks and Modbus.

Characteristics

Building management systems are most commonly implemented in large projects with extensive mechanical, HVAC, electrical systems. Systems linked to a BMS typically represent 40% of a building's energy usage; if lighting is included, this number approaches to 70%. BMS systems are a critical component to managing energy demand. Improperly configured BMS systems are believed to account for 20% of building energy usage, or approximately 8% of total energy usage in the United States.

In addition to controlling the building's internal environment, BMS systems are sometimes linked to access control (turnstiles and access doors controlling who is allowed access and egress to the building) or other security systems such as closed-circuit television (CCTV) and motion detectors. Fire alarm systems and elevators are also sometimes linked to a BMS, for monitoring. In case a fire is detected then only the fire alarm panel could shut off dampers in the ventilation system to stop smoke spreading and send all the elevators to the ground floor and park them to prevent people from using them.

- Illumination (lighting) control
- Electric power control
- Heating, ventilation and air-conditioning (HVAC)
- Security and observation
- Access control
- Fire alarm system
- Lifts, elevators etc.
- Plumbing
- Closed-circuit television (CCTV)
- Other engineering systems
- Control Panel
- PA system
- Alarm Monitor
- Security Automation

Benefits

- Possibility of individual room control

- Increased staff productivity

- Effective monitoring and targeting of energy consumption

- Improved plant reliability and life

- Effective response to HVAC-related complaints

- Save time and money during the maintenance.

Building Managers

- Higher rental value

- Flexibility on change of building use

- Individual tenant billing for services facilities time saving

- Remote monitoring of the plants (such as AHU's, fire pumps, plumbing pumps, electrical supply, STP, WTP, grey water treatment plant etc.)

Maintenance Companies

- Ease of information availability

- Computerized maintenance scheduling

- Effective use of maintenance staff

- Early detection of problems or service work easy

- More satisfied occupants

Additional Benefits

- Data is consolidated onto a single system to improve reporting, information management and decision-making. Integrating and managing the HVAC, energy, security, digital video and life safety applications from a single workstation allows facility-wide insight and control for better performance.

- Increased operational savings – Efficient resource deployment can result in reduced operational costs, empowering operators, simplifying training and decreasing false alarms.

- Energy efficient – Real-time view into facility operations and deep trend analysis provide data-driven insight to optimize your energy management strategies and minimize operational costs.

- Flexibility to grow and expand – The powerful combination of open systems protocols and a scalable platform means the BMS can help support growth and expansion of the system in the future.

- Reduced risk – Strategic mobile or desktop control, exceptional alarm management and integrated security solutions helps to see the big picture, helping to speed up response time and mitigate risks for the property, people and business.

- Intelligent reporting – Comprehensive reporting with functionality for customizable reports delivers greater transparency into system history and promotes compliance.

BMS deals with energy demand management. EDM integrates energy policies and regulations in to overall company operations. It incorporates energy targets into overall business strategies. EDM conduct management reviews and establishes a system to collect, analyse and report data related energy consumption and ensure correctness and integrity of that data.

Hierarchical Control System

A hierarchical control system is a form of control system in which a set of devices and governing software is arranged in a hierarchical tree. When the links in the tree are implemented by a computer network, then that hierarchical control system is also a form of networked control system.

Hierarchical Control System

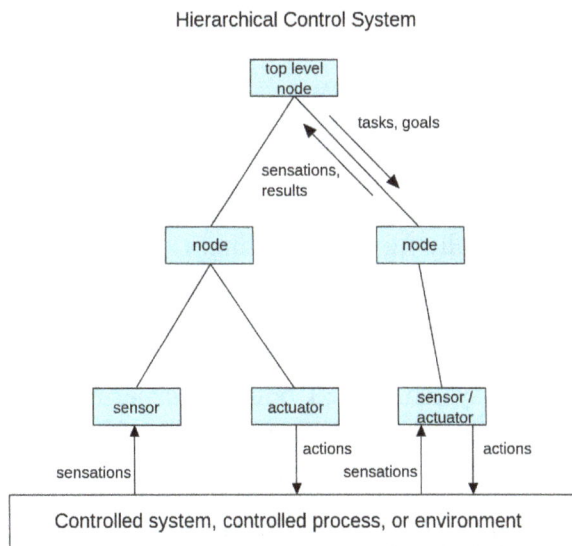

A hierarchical control system takes the shape of a tree in which each node operates independently, performing tasks from its superior node, commanding tasks of its subordinate nodes, sending abstracted sensations to its superior node, and receiving sensations from its subordinate nodes. Leaf nodes are sensors or actuators.

Overview

A human-built system with complex behavior is often organized as a hierarchy. For example, a command hierarchy has among its notable features the organizational chart of superiors, subordinates, and lines of organizational communication. Hierarchical control systems are organized similarly to divide the decision making responsibility.

Each element of the hierarchy is a linked node in the tree. Commands, tasks and goals to be achieved flow down the tree from superior nodes to subordinate nodes, whereas sensations and command results flow up the tree from subordinate to superior nodes. Nodes may also exchange messages with their siblings. The two distinguishing features of a hierarchical control system are related to its layers.

- Each higher layer of the tree operates with a longer interval of planning and execution time than its immediately lower layer.

- The lower layers have local tasks, goals, and sensations, and their activities are planned and coordinated by higher layers which do not generally override their decisions. The layers form a hybrid intelligent system in which the lowest, reactive layers are sub-symbolic. The higher layers, having relaxed time constraints, are capable of reasoning from an abstract world model and performing planning. A hierarchical task network is a good fit for planning in a hierarchical control system.

Besides artificial systems, an animal's control systems are proposed to be organized as a hierarchy. In perceptual control theory, which postulates that an organism's behavior is a means of controlling its perceptions, the organism's control systems are suggested to be organized in a hierarchical pattern as their perceptions are constructed so.

Applications

Manufacturing, Robotics and Vehicles

Among the robotic paradigms is the hierarchical paradigm in which a robot operates in a top-down fashion, heavy on planning, especially motion planning. Computer-aided production engineering has been a research focus at NIST since the 1980s. Its Automated Manufacturing Research Facility was used to develop a five layer production control model. In the early 1990s DARPA sponsored research to develop distributed (i.e. networked) intelligent control systems for applications such as military command and control systems. NIST built on earlier research to develop its Real-Time Control System (RCS) and Real-time Control System Software which is a generic hierarchical control system that has been used to operate a manufacturing cell, a robot crane, and an automated vehicle.

In November 2007, DARPA held the Urban Challenge. The winning entry, Tartan Racing employed a hierarchical control system, with layered mission planning, motion planning, behavior generation, perception, world modelling, and mechatronics.

Artificial Intelligence

Subsumption architecture is a methodology for developing artificial intelligence that is heavily associated with behavior based robotics. This architecture is a way of decomposing complicated intelligent behavior into many "simple" behavior modules, which are in turn organized into layers. Each layer implements a particular goal of the software agent (i.e. system as a whole), and higher layers are increasingly more abstract. Each layer's goal subsumes that of the underlying layers, e.g. the decision to move forward by the eat-food layer takes into account the decision of the lowest obstacle-avoidance layer. Behavior need not be planned by a superior layer, rather behaviors may be triggered by sensory inputs and so are only active under circumstances where they might be appropriate.

Reinforcement learning has been used to acquire behavior in a hierarchical control system in which each node can learn to improve its behavior with experience.

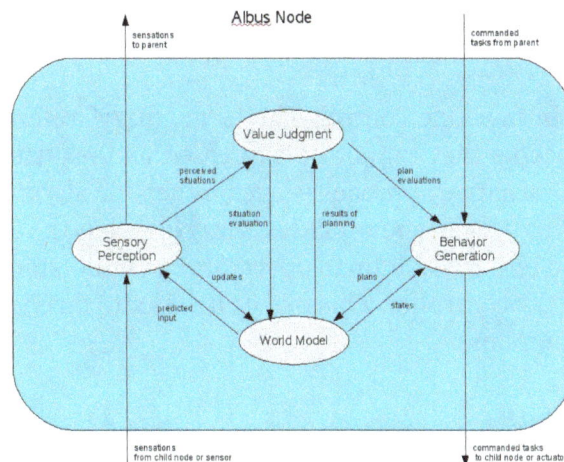

Constituents in a node from James Albus's Reference Model Architecture

James Albus, while at NIST, developed a theory for intelligent system design named the Reference Model Architecture (RMA), which is a hierarchical control system inspired by RCS. Albus defines each node to contain these components.

- *Behavior generation* is responsible for executing tasks received from the superior, parent node. It also plans for, and issues tasks to, the subordinate nodes.

- *Sensory perception* is responsible for receiving sensations from the subordinate nodes, then grouping, filtering, and otherwise processing them into higher level abstractions that update the local state and which form sensations that are sent to the superior node.

- *Value judgment* is responsible for evaluating the updated situation and evaluating alternative plans.

- *World Model* is the local state that provides a model for the controlled system, controlled process, or environment at the abstraction level of the subordinate nodes.

At its lowest levels, the RMA can be implemented as a subsumption architecture, in which the world model is mapped directly to the controlled process or real world, avoiding the need for a mathematical abstraction, and in which time-constrained reactive planning can be implemented as a finite state machine. Higher levels of the RMA however, may have sophisticated mathematical world models and behavior implemented by automated planning and scheduling. Planning is required when certain behaviors cannot be triggered by current sensations, but rather by predicted or anticipated sensations, especially those that come about as result of the node's actions.

Fuzzy Control System

A fuzzy control system is a control system based on fuzzy logic—a mathematical system that analyzes analog input values in terms of logical variables that take on continuous values between 0 and 1, in contrast to classical or digital logic, which operates on discrete values of either 1 or 0 (true or false, respectively).

Overview

Fuzzy logic is widely used in machine control. The term "fuzzy" refers to the fact that the logic involved can deal with concepts that cannot be expressed as the "true" or "false" but rather as "partially true". Although alternative approaches such as genetic algorithms and neural networks can perform just as well as fuzzy logic in many cases, fuzzy logic has the advantage that the solution to the problem can be cast in terms that human operators can understand, so that their experience can be used in the design of the controller. This makes it easier to mechanize tasks that are already successfully performed by humans.

History and Applications

Fuzzy logic was first proposed by Lotfi A. Zadeh of the University of California at Berkeley in a 1965 paper. He elaborated on his ideas in a 1973 paper that introduced the concept of "linguistic variables", which in this article equates to a variable defined as a fuzzy set. Other research followed, with the first industrial application, a cement kiln built in Denmark, coming on line in 1975.

Fuzzy systems were initially implemented in Japan.

- Interest in fuzzy systems was sparked by Seiji Yasunobu and Soji Miyamoto of Hitachi, who in 1985 provided simulations that demonstrated the feasibility of fuzzy control systems for the Sendai railway. Their ideas were adopted, and fuzzy systems were used to control accelerating, braking, and stopping when the line opened in 1987.

- In 1987, Takeshi Yamakawa demonstrated the use of fuzzy control, through a set of simple dedicated fuzzy logic chips, in an "inverted pendulum" experiment. This is a classic control problem, in which a vehicle tries to keep a pole mounted on its top by a hinge upright by moving back and forth. Yamakawa subsequently made the demonstration more sophisticated by mounting a wine glass containing water and even a live mouse to the top of the pendulum: the system maintained stability in both cases. Yamakawa eventually went on to organize his own fuzzy-systems research lab to help exploit his patents in the field.

- Japanese engineers subsequently developed a wide range of fuzzy systems for both industrial and consumer applications. In 1988 Japan established the Laboratory for International Fuzzy Engineering (LIFE), a cooperative arrangement between 48 companies to pursue fuzzy research. The automotive company Volkswagen was the only foreign corporate member of LIFE, dispatching a researcher for a duration of three years.

- Japanese consumer goods often incorporate fuzzy systems. Matsushita vacuum cleaners use microcontrollers running fuzzy algorithms to interrogate dust sensors and adjust suction power accordingly. Hitachi washing machines use fuzzy controllers to load-weight, fabric-mix, and dirt sensors and automatically set the wash cycle for the best use of power, water, and detergent.

- Canon developed an autofocusing camera that uses a charge-coupled device (CCD) to measure the clarity of the image in six regions of its field of view and use the information provided to determine if the image is in focus. It also tracks the rate of change of lens movement during focusing, and controls its speed to prevent overshoot. The camera's fuzzy control system uses 12 inputs: 6 to obtain the current clarity data provided by the CCD and 6 to measure the rate of change of lens movement. The output is the position of the lens. The fuzzy control system uses 13 rules and requires 1.1 kilobytes of memory.

- An industrial air conditioner designed by Mitsubishi uses 25 heating rules and 25 cooling rules. A temperature sensor provides input, with control outputs fed to an inverter, a compressor valve, and a fan motor. Compared to the previous design, the fuzzy controller heats and cools five times faster, reduces power consumption by 24%, increases temperature stability by a factor of two, and uses fewer sensors.

- Other applications investigated or implemented include: character and handwriting recognition; optical fuzzy systems; robots, including one for making Japanese flower arrangements; voice-controlled robot helicopters (hovering is a "balancing act" rather similar to the inverted pendulum problem); rehabilitation robotics to provide patient-specific solutions (e.g. to control heart rate and blood pressure); control of flow of powders in film manufacture; elevator systems; and so on.

Work on fuzzy systems is also proceeding in the United State and Europe, although on a less extensive scale than in Japan.

- The US Environmental Protection Agency has investigated fuzzy control for energy-efficient motors, and NASA has studied fuzzy control for automated space docking: simulations show that a fuzzy control system can greatly reduce fuel consumption.

- Firms such as Boeing, General Motors, Allen-Bradley, Chrysler, Eaton, and Whirlpool have worked on fuzzy logic for use in low-power refrigerators, improved automotive transmissions, and energy-efficient electric motors.

- In 1995 Maytag introduced an "intelligent" dishwasher based on a fuzzy controller and a "one-stop sensing module" that combines a thermistor, for temperature measurement; a conductivity sensor, to measure detergent level from the ions present in the wash; a turbidity sensor that measures scattered and transmitted light to measure the soiling of the wash; and a magnetostrictive sensor to read spin rate. The system determines the optimum wash cycle for any load to obtain the best results with the least amount of energy, detergent, and water. It even adjusts for dried-on foods by tracking the last time the door was opened, and estimates the number of dishes by the number of times the door was opened.

Research and development is also continuing on fuzzy applications in software, as opposed to firmware, design, including fuzzy expert systems and integration of fuzzy logic with neural-network and so-called adaptive "genetic" software systems, with the ultimate goal of building "self-learning" fuzzy-control systems. These systems can be employed to control complex, nonlinear dynamic plants, for example, human body.

Fuzzy Sets

The input variables in a fuzzy control system are in general mapped by sets of membership functions similar to this, known as "fuzzy sets". The process of converting a crisp input value to a fuzzy value is called "fuzzification".

A control system may also have various types of switch, or "ON-OFF", inputs along with its analog inputs, and such switch inputs of course will always have a truth value equal to either 1 or 0, but the scheme can deal with them as simplified fuzzy functions that happen to be either one value or another.

Given "mappings" of input variables into membership functions and truth values, the microcontroller then makes decisions for what action to take, based on a set of "rules", each of the form:

IF brake temperature IS warm AND speed IS not very fast

THEN brake pressure IS slightly decreased.

In this example, the two input variables are "brake temperature" and "speed" that have values defined as fuzzy sets. The output variable, "brake pressure" is also defined by a fuzzy set that can have values like "static" or "slightly increased" or "slightly decreased" etc.

This rule by itself is very puzzling since it looks like it could be used without bothering with fuzzy logic, but remember that the decision is based on a set of rules:

- All the rules that apply are invoked, using the membership functions and truth values obtained from the inputs, to determine the result of the rule.

- This result in turn will be mapped into a membership function and truth value controlling the output variable.

- These results are combined to give a specific ("crisp") answer, the actual brake pressure, a procedure known as "defuzzification".

This combination of fuzzy operations and rule-based "inference" describes a "fuzzy expert system".

Traditional control systems are based on mathematical models in which the control system is described using one or more differential equations that define the system response to its inputs. Such systems are often implemented as "PID controllers" (proportional-integral-derivative controllers). They are the products of decades of development and theoretical analysis, and are highly effective.

If PID and other traditional control systems are so well-developed, why bother with fuzzy control? It has some advantages. In many cases, the mathematical model of the control process may not exist, or may be too "expensive" in terms of computer processing power and memory, and a system based on empirical rules may be more effective.

Furthermore, fuzzy logic is well suited to low-cost implementations based on cheap sensors, low-resolution analog-to-digital converters, and 4-bit or 8-bit one-chip microcontroller chips. Such systems can be easily upgraded by adding new rules to improve performance or add new features. In many cases, fuzzy control can be used to improve existing traditional controller systems by adding an extra layer of intelligence to the current control method.

Fuzzy Control in Detail

Fuzzy controllers are very simple conceptually. They consist of an input stage, a processing stage, and an output stage. The input stage maps sensor or other inputs, such as switches, thumbwheels, and so on, to the appropriate membership functions and truth values. The processing stage invokes each appropriate rule and generates a result for each, then combines the results of the rules. Finally, the output stage converts the combined result back into a specific control output value.

The most common shape of membership functions is triangular, although trapezoidal and bell curves are also used, but the shape is generally less important than the number of curves and their placement. From three to seven curves are generally appropriate to cover the required range of an input value, or the "universe of discourse" in fuzzy jargon.

As discussed earlier, the processing stage is based on a collection of logic rules in the form of IF-THEN statements, where the IF part is called the "antecedent" and the THEN part is called the "consequent". Typical fuzzy control systems have dozens of rules.

Consider a rule for a thermostat:

IF (temperature is "cold") THEN (heater is "high")

This rule uses the truth value of the "temperature" input, which is some truth value of "cold", to generate a result in the fuzzy set for the "heater" output, which is some value of "high". This result is used with the results of other rules to finally generate the crisp composite output. Obviously, the greater the truth value of "cold", the higher the truth value of "high", though this does not necessarily mean that the output itself will be set to "high" since this is only one rule among many. In some cases, the membership functions can be modified by "hedges" that are equivalent to adverbs. Common hedges include "about", "near", "close to", "approximately", "very", "slightly", "too", "extremely", and "somewhat". These operations may have precise definitions, though the definitions can vary considerably between different implementations. "Very", for one example, squares membership functions; since the membership values are always less than 1, this narrows the membership function. "Extremely" cubes the values to give greater narrowing, while "somewhat" broadens the function by taking the square root.

In practice, the fuzzy rule sets usually have several antecedents that are combined using fuzzy operators, such as AND, OR, and NOT, though again the definitions tend to vary: AND, in one popular definition, simply uses the minimum weight of all the antecedents, while OR uses the maximum value. There is also a NOT operator that subtracts a membership function from 1 to give the "complementary" function.

There are several ways to define the result of a rule, but one of the most common and simplest is the "max-min" inference method, in which the output membership function is given the truth value generated by the premise.

Rules can be solved in parallel in hardware, or sequentially in software. The results of all the rules that have fired are "defuzzified" to a crisp value by one of several methods. There are dozens, in theory, each with various advantages or drawbacks.

The "centroid" method is very popular, in which the "center of mass" of the result provides the crisp value. Another approach is the "height" method, which takes the value of the biggest contributor. The centroid method favors the rule with the output of greatest area, while the height method obviously favors the rule with the greatest output value.

The diagram below demonstrates max-min inferencing and centroid defuzzification for a system with input variables "x", "y", and "z" and an output variable "n". Note that "mu" is standard fuzzy-logic nomenclature for "truth value":

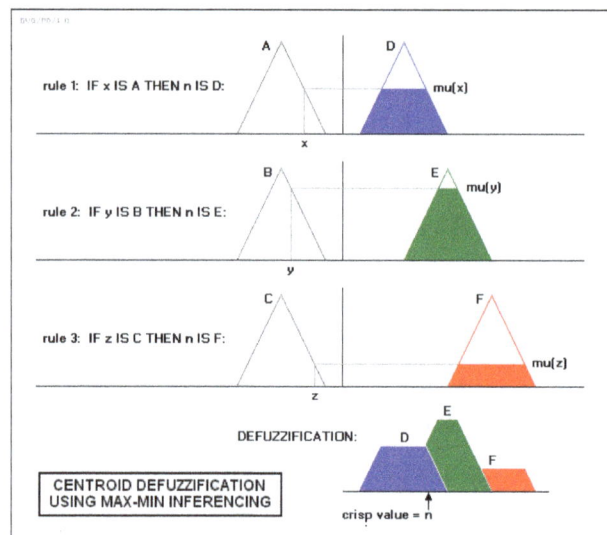

Notice how each rule provides a result as a truth value of a particular membership function for the output variable. In centroid defuzzification the values are OR'd, that is, the maximum value is used and values are not added, and the results are then combined using a centroid calculation.

Fuzzy control system design is based on empirical methods, basically a methodical approach to trial-and-error. The general process is as follows:

- Document the system's operational specifications and inputs and outputs.

- Document the fuzzy sets for the inputs.

- Document the rule set.

- Determine the defuzzification method.

- Run through test suite to validate system, adjust details as required.

- Complete document and release to production.

As a general example, consider the design of a fuzzy controller for a steam turbine. The block diagram of this control system appears as follows:

The input and output variables map into the following fuzzy set:

—where:

N3:   Large negative.

N2:   Medium negative.

N1:   Small negative.

Z: Zero.

P1:   Small positive.

P2:   Medium positive.

P3:   Large positive.

The rule set includes such rules as:

rule 1:   IF temperature IS cool AND pressure IS weak,

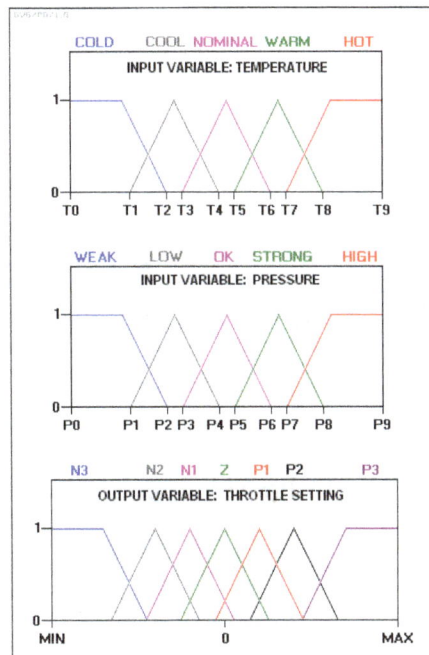

       THEN throttle is P3.

rule 2: IF temperature IS cool AND pressure IS low,

       THEN throttle is P2.

rule 3: IF temperature IS cool AND pressure IS ok,

       THEN throttle is Z.

rule 4: IF temperature IS cool AND pressure IS strong,

       THEN throttle is N2.

In practice, the controller accepts the inputs and maps them into their membership functions and truth values. These mappings are then fed into the rules. If the rule specifies an AND relationship between the mappings of the two input variables, as the examples above do, the minimum of the two is used as the combined truth value; if an OR is specified, the maximum is used. The appropriate output state is selected and assigned a membership value at the truth level of the premise. The truth values are then defuzzified. For an example, assume the temperature is in the "cool" state, and the pressure is in the "low" and "ok" states. The pressure values ensure that only rules 2 and 3 fire:

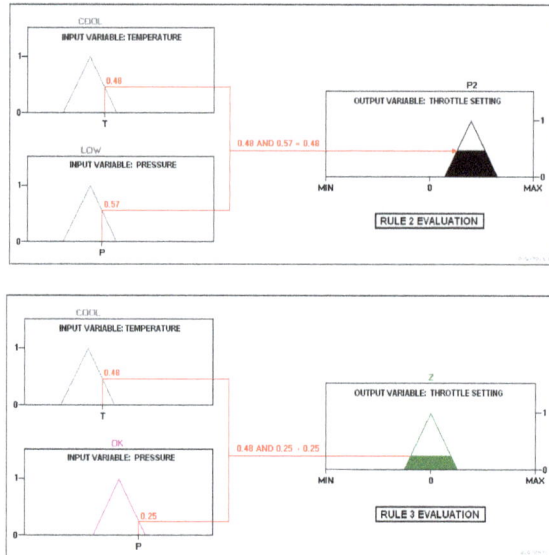

The two outputs are then defuzzified through centroid defuzzification:

The output value will adjust the throttle and then the control cycle will begin again to generate the next value .

Building a Fuzzy Controller

Consider implementing with a microcontroller chip a simple feedback controller:

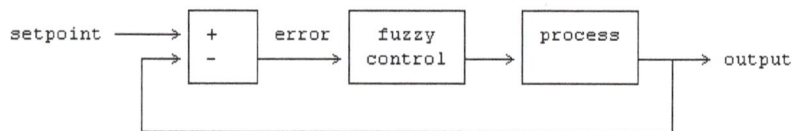

A fuzzy set is defined for the input error variable "e", and the derived change in error, "delta", as well as the "output", as follows:

LP: large positive

SP: small positive

ZE: zero

SN: small negative

LN: large negative

If the error ranges from -1 to +1, with the analog-to-digital converter used having a resolution of 0.25, then the input variable's fuzzy set (which, in this case, also applies to the output variable) can be described very simply as a table, with the error / delta / output values in the top row and the truth values for each membership function arranged in rows beneath:

	-1	-0.75	-0.5	-0.25	0	0.25	0.5	0.75	1
mu(LP)	0	0	0	0	0	0	0.3	0.7	1
mu(SP)	0	0	0	0	0.3	0.7	1	0.7	0.3
mu(ZE)	0	0	0.3	0.7	1	0.7	0.3	0	0
mu(SN)	0.3	0.7	1	0.7	0.3	0	0	0	0
mu(LN)	1	0.7	0.3	0	0	0	0	0	0

—or, in graphical form (where each "X" has a value of 0.1):

```
        LN              SN              ZE              SP              LP
      r - - - - - - - - - - - - - - - - - - - - - - - - - - - - - - - - - - - 1
-1.0  | XXXXXXXXXX    XXX             :               :               :       |
      |                                                                       |
-0.75 | XXXXXXX       XXXXXXX         :               :               :       |
      |                                                                       |
-0.5  | XXX           XXXXXXXXXX    XXX               :               :       |
      |                                                                       |
-0.25 | :            XXXXXXX         XXXXXXX          :               :       |
      |                                                                       |
0.0   | :            XXX             XXXXXXXXXX    XXX                 :       |
      |                                                                       |
0.25  | :             :              XXXXXXX         XXXXXXX           :       |
      |                                                                       |
0.5   | :             :              XXX             XXXXXXXXXX    XXX         |
      |                                                                       |
0.75  | :             :               :              XXXXXXX         XXXXXXX   |
      |                                                                       |
1.0   | :             :               :              XXX             XXXXXXXXXX|
      |                                                                       |
      L - - - - - - - - - - - - - - - - - - - - - - - - - - - - - - - - - - - J
```

Suppose this fuzzy system has the following rule base:

  rule 1: IF e = ZE AND delta = ZE THEN output = ZE

  rule 2: IF e = ZE AND delta = SP THEN output = SN

  rule 3: IF e = SN AND delta = SN THEN output = LP

  rule 4: IF e = LP OR delta = LP THEN output = LN

These rules are typical for control applications in that the antecedents consist of the logical combination of the error and error-delta signals, while the consequent is a control command output. The rule outputs can be defuzzified using a discrete centroid computation:

SUM(I = 1 TO 4 OF (mu(I) * output(I))) / SUM(I = 1 TO 4 OF mu(I))

Now, suppose that at a given time we have:

  e = 0.25

  delta = 0.5

Then this gives:

	e	delta
mu(LP)	0	0.3
mu(SP)	0.7	1
mu(ZE)	0.7	0.3
mu(SN)	0	0
mu(LN)	0	0

Plugging this into rule 1 gives:

rule 1: IF e = ZE AND delta = ZE THEN output = ZE

   mu(1) = MIN(0.7, 0.3) = 0.3

output(1) = 0

-- where:

- mu(1): Truth value of the result membership function for rule 1. In terms of a centroid calculation, this is the "mass" of this result for this discrete case.

- output(1): Value (for rule 1) where the result membership function (ZE) is maximum over the output variable fuzzy set range. That is, in terms of a centroid calculation, the location of the "center of mass" for this individual result. This value is independent of the value of "mu". It simply identifies the location of ZE along the output range.

The other rules give:

rule 2: IF e = ZE AND delta = SP THEN output = SN

   mu(2) = MIN(0.7, 1) = 0.7

output(2) = -0.5

rule 3: IF e = SN AND delta = SN THEN output = LP

   mu(3) = MIN(0.0, 0.0) = 0

output(3) = 1

rule 4: IF e = LP OR delta = LP THEN output = LN

   mu(4) = MAX(0.0, 0.3) = 0.3

output(4) = -1

The centroid computation yields:

$$\frac{mu(1).output(1) + mu(2).output(2) + mu(3).output(3) + mu(4).output(4)}{mu(1) + mu(2) + mu(3) + mu(4)}$$

$$= \frac{(0.3*0)+(0.7*-0.5)+(0*1)+(0.3*-1)}{0.3+0.7+0+0.3}$$

—for the final control output. Simple. Of course the hard part is figuring out what rules actually work correctly in practice.

If you have problems figuring out the centroid equation, remember that a centroid is defined by summing all the moments (location times mass) around the center of gravity and equating the sum to zero. So if X_0 is the center of gravity, X_i is the location of each mass, and M_i is each mass, this gives:

$$0 = (X_1 - X_0)*M_1 + (X_2 - X_0)*M_2 + \ldots + (X_n - X_0)*M_n$$

$$0 = (X_1*M_1 + X_2*M_2 + \ldots + X_n*M_n) - X_0*(M_1 + M_2 + \ldots + M_n)$$

$$X_0*(M_1 + M_2 + \ldots + M_n) = X_1*M_1 + X_2*M_2 + \ldots + X_n*M_n$$

$$X_0 = \frac{X_1*M_1 + X_2*M_2 + \ldots + X_n*M_n}{M_1 + M_2 + \ldots + M_n}$$

In our example, the values of mu correspond to the masses, and the values of X to location of the masses (mu, however, only 'corresponds to the masses' if the initial 'mass' of the output functions are all the same/equivalent. If they are not the same, i.e. some are narrow triangles, while others maybe wide trapizoids or shouldered triangles, then the mass or area of the output function must be known or calculated. It is this mass that is then scaled by mu and multiplied by its location X_i).

This system can be implemented on a standard microprocessor, but dedicated fuzzy chips are now available. For example, Adaptive Logic INC of San Jose, California, sells a "fuzzy chip", the AL220, that can accept four analog inputs and generate four analog outputs.

Antilock Brakes

As a first example, consider an anti-lock braking system, directed by a microcontroller chip. The microcontroller has to make decisions based on brake temperature, speed, and other variables in the system.

The variable "temperature" in this system can be subdivided into a range of "states": "cold", "cool", "moderate", "warm", "hot", "very hot". The transition from one state to the next is hard to define.

An arbitrary static threshold might be set to divide "warm" from "hot". For example, at exactly 90 degrees, warm ends and hot begins. But this would result in a discontinuous change when the input value passed over that threshold. The transition wouldn't be smooth, as would be required in braking situations.

The way around this is to make the states *fuzzy*. That is, allow them to change gradually from one state to the next. In order to do this there must be a dynamic relationship established between different factors.

We start by defining the input temperature states using "membership functions":

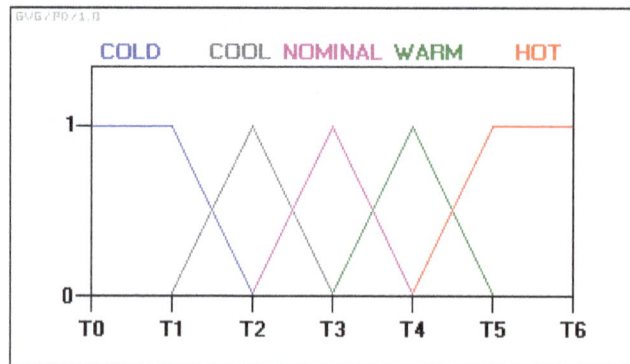

With this scheme, the input variable's state no longer jumps abruptly from one state to the next. Instead, as the temperature changes, it loses value in one membership function while gaining value in the next. In other words, its ranking in the category of cold decreases as it becomes more highly ranked in the warmer category.

At any sampled timeframe, the "truth value" of the brake temperature will almost always be in some degree part of two membership functions: i.e.: '0.6 nominal and 0.4 warm', or '0.7 nominal and 0.3 cool', and so on.

The above example demonstrates a simple application, using the abstraction of values from multiple values. This only represents one kind of data, however, in this case, temperature.

Adding additional sophistication to this braking system, could be done by additional factors such as traction, speed, inertia, set up in dynamic functions, according to the designed fuzzy system.

Logical Interpretation of Fuzzy Control

In spite of the appearance there are several difficulties to give a rigorous logical interpretation of the *IF-THEN* rules. As an example, interpret a rule as *IF (temperature is "cold") THEN (heater is "high")* by the first order formula *Cold(x)→High(y)* and assume that r is an input such that *Cold(r)* is false. Then the formula *Cold(r)→High(t)* is true for any *t* and therefore any *t* gives a correct control given *r*. A rigorous logical justification of fuzzy control is given in Hájek's book where fuzzy control is represented as a theory of Hájek's basic logic. Also in Gerla 2005 another logical approach to fuzzy control is proposed based on fuzzy logic programming.Indeed, denote by *f* the fuzzy function arising of an IF-THEN systems of rules. Then we can translate this system into a fuzzy program P containing a series of rules whose head is "Good(x,y)". The interpretation of this predicate in the least fuzzy Herbrand model of P coincides with f. This gives further useful tools to fuzzy control.

Real-time Control System

Real-time Control System (RCS) is a reference model architecture, suitable for many software-intensive, real-time control problem domains. RCS is a reference model architecture that defines the types of functions that are required in a real-time intelligent control system, and how these functions are related to each other.

RCS is not a system design, nor is it a specification of how to implement specific systems. RCS prescribes a hierarchical control model based on a set of well-founded engineering principles to organize system complexity. All the control nodes at all levels share a generic node model.

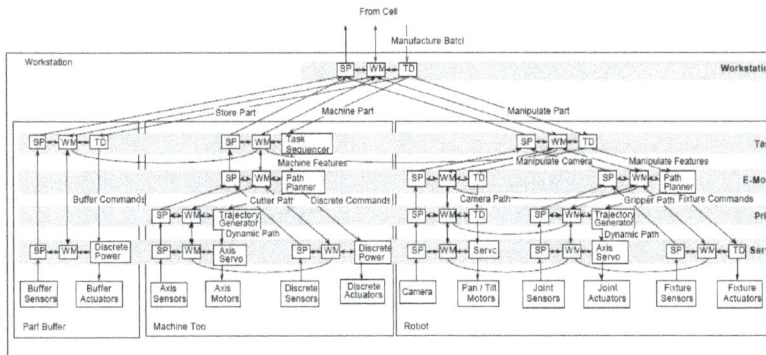

Example of a RCS-3 application of a machining workstation containing a machine tool, part buffer, and robot with vision system. RCS-3 produces a layered graph of processing nodes, each of which contains a task decomposition (TD), world modeling (WM), and sensory processing (SP) module. These modules are richly interconnected to each other by a communications system.

Also RCS provides a comprehensive methodology for designing, engineering, integrating, and testing control systems. Architects iteratively partition system tasks and information into finer, finite subsets that are controllable and efficient. RCS focuses on intelligent control that adapts to uncertain and unstructured operating environments. The key concerns are sensing, perception, knowledge, costs, learning, planning, and execution.

Overview

A reference model architecture is a canonical form, not a system design specification. The RCS reference model architecture combines real-time motion planning and control with high level task planning, problem solving, world modeling, recursive state estimation, tactile and visual image processing, and acoustic signature analysis. In fact, the evolution of the RCS concept has been driven by an effort to include the best properties and capabilities of most, if not all, the intelligent control systems currently known in the literature, from subsumption to SOAR, from blackboards to object-oriented programming.

RCS (real-time control system) is developed into an intelligent agent architecture designed to enable any level of intelligent behavior, up to and including human levels of performance. RCS was inspired by a theoretical model of the cerebellum, the portion of the brain responsible for fine motor coordination and control of conscious motions. It was originally designed for sensory-interactive goal-directed control of laboratory manipulators. Over three decades, it has evolved into a real-time control architecture for intelligent machine tools, factory automation systems, and intelligent autonomous vehicles.

RCS applies to many problem domains including manufacturing examples and vehicle systems examples. Systems based on the RCS architecture have been designed and implemented to varying degrees for a wide variety of applications that include loading and unloading of parts and tools in machine tools, controlling machining workstations, performing robotic deburring and chamfering, and controlling space station telerobots, multiple autonomous undersea vehicles, unmanned land vehicles, coal mining automation systems, postal service mail handling systems, and submarine operational automation systems.

History

RCS has evolved through a variety of versions over a number of years as understanding of the complexity and sophistication of intelligent behavior has increased. The first implementation was designed for sensory-interactive robotics by Barbera in the mid 1970s.

RCS-1

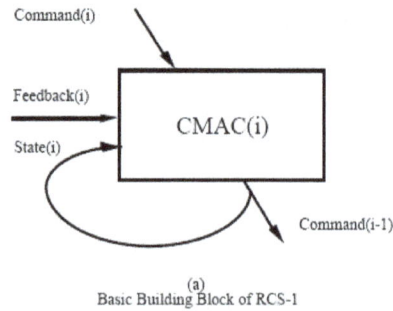

(a)
Basic Building Block of RCS-1

Basics of the RCS-1 control paradigm.

In RCS-1, the emphasis was on combining commands with sensory feedback so as to compute the proper response to every combination of goals and states. The application was to control a robot arm with a structured light vision system in visual pursuit tasks. RCS-1 was heavily influenced by biological models such as the Marr-Albus model, and the Cerebellar Model Arithmetic Computer (CMAC). of the cerebellum.

CMAC becomes a state machine when some of its outputs are fed directly back to the input, so RCS-1 was implemented as a set of state-machines arranged in a hierarchy of control levels. At each level, the input command effectively selects a behavior that is driven by feedback in stimulus-response fashion. CMAC thus became the reference model building block of RCS-1, as shown in the figure.

A hierarchy of these building blocks was used to implement a hierarchy of behaviors such as observed by Tinbergen and others. RCS-1 is similar in many respects to Brooks' subsumption architecture, except that RCS selects behaviors before the fact through goals expressed in commands, rather than after the fact through subsumption.

RCS-2

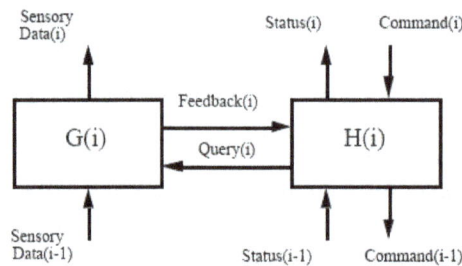

RCS-2 control paradigm.

The next generation, RCS-2, was developed by Barbera, Fitzgerald, Kent, and others for manufacturing control in the NIST Automated Manufacturing Research Facility (AMRF) during the early 1980s. The basic building block of RCS-2 is shown in the figure.

The H function remained a finite state machine state-table executor. The new feature of RCS-2 was the inclusion of the G function consisting of a number of sensory processing algorithms including structured light and blob analysis algorithms. RCS-2 was used to define an eight level hierarchy consisting of Servo, Coordinate Transform, E-Move, Task, Workstation, Cell, Shop, and Facility levels of control.

Only the first six levels were actually built. Two of the AMRF workstations fully implemented five levels of RCS-2. The control system for the Army Field Material Handling Robot (FMR) was also implemented in RCS-2, as was the Army TMAP semi-autonomous land vehicle project.

RCS-3

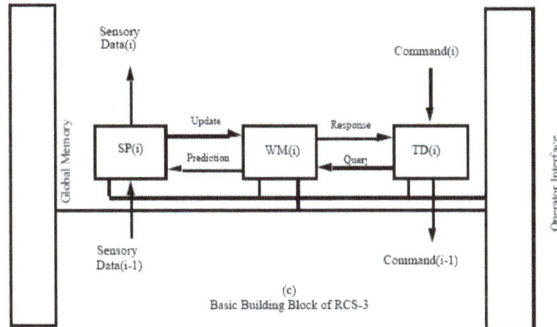

RCS-3 control paradigm.

RCS-3 was designed for the NBS/DARPA Multiple Autonomous Undersea Vehicle (MAUV) project and was adapted for the NASA/NBS Standard Reference Model Telerobot Control System Architecture (NASREM) developed for the space station Flight Telerobotic Servicer The basic building block of RCS-3 is shown in the figure.

The principal new features introduced in RCS-3 are the World Model and the operator interface. The inclusion of the World Model provides the basis for task planning and for model-based sensory processing. This led to refinement of the task decomposition (TD) modules so that each have a job assigner, and planner and executor for each of the subsystems assigned a job. This corresponds roughly to Saridis' three level control hierarchy.

RCS-4

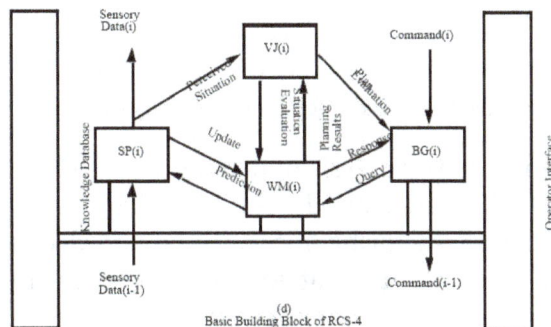

RCS-4 control paradigm.

RCS-4 is developed since the 1990s by the NIST Robot Systems Division. The basic building block is shown in the figure). The principal new feature in RCS-4 is the explicit representation of the Value Judgment (VJ) system. VJ modules provide to the RCS-4 control system the type of functions provided to the biological brain by the limbic system. The VJ modules contain processes that compute cost, benefit, and risk of planned actions, and that place value on objects, materials, territory, situations, events, and outcomes. Value state-variables define what goals are important and what objects or regions should be attended to, attacked, defended, assisted, or otherwise acted upon. Value judgments, or evaluation functions, are an essential part of any form of planning or learning. The application of value judgments to intelligent control systems has been addressed by George Pugh. The structure and function of VJ modules are developed more completely developed in Albus (1991).

RCS-4 also uses the term behavior generation (BG) in place of the RCS-3 term task 5 decomposition (TD). The purpose of this change is to emphasize the degree of autonomous decision making. RCS-4 is designed to address highly autonomous applications in unstructured environments where high bandwidth communications are impossible, such as unmanned vehicles operating on the battlefield, deep undersea, or on distant planets. These applications require autonomous value judgments and sophisticated real-time perceptual capabilities. RCS-3 will continue to be used for less demanding applications, such as manufacturing, construction, or telerobotics for near-space, or shallow undersea operations, where environments are more structured and communication bandwidth to a human interface is less restricted. In these applications, value judgments are often represented implicitly in task planning processes, or in human operator input.

RCS Methodology

In the figure, an example of the RCS methodology for designing a control system for autonomous onroad driving under everyday traffic conditions is summarized in six steps.

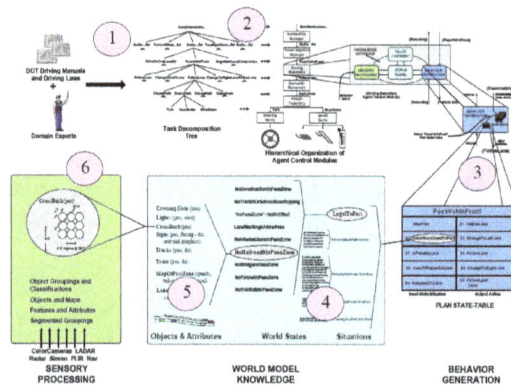

The six steps of the RCS methodology for knowledge acquisition and representation.

- Step 1 consists of an intensive analysis of domain knowledge from training manuals and subject matter experts. Scenarios are developed and analyzed for each task and subtask. The result of this step is a structuring of procedural knowledge into a task decomposition tree with simpler and simpler tasks at each echelon. At each echelon, a vocabulary of commands (action verbs with goal states, parameters, and constraints) is defined to evoke task behavior at each echelon.

- Step 2 defines a hierarchical structure of organizational units that will execute the commands defined in step 1. For each unit, its duties and responsibilities in response to each command are specified. This is analogous to establishing a work breakdown structure for a development project, or defining an organizational chart for a business or military operation.

- Step 3 specifies the processing that is triggered within each unit upon receipt of an input command. For each input command, a state-graph (or statetable or extended finite state automaton) is defined that provides a plan (or procedure for making a plan) for accomplishing the commanded task. The input command selects (or causes to be generated) an appropriate state-table, the execution of which generates a series of output commands to units at the next lower echelon. The library of state-tables contains a set of statesensitive procedural rules that identify all the task branching conditions and specify the corresponding state transition and output command parameters.

The result of step 3 is that each organizational unit has for each input command a state-table of ordered production rules, each suitable for execution by an extended finite state automaton (FSA). The sequence of output subcommands required to accomplish the input command is generated by situations (i.e., branching conditions) that cause the FSA to transition from one output subcommand to the next.

- In step 4, each of the situations that are defined in step 3 are analyzed to reveal their dependencies on world and task states. This step identifies the detailed relationships between entities, events, and states of the world that cause a particular situation to be true.

- In step 5, we identify and name all of the objects and entities together with their particular features and attributes that are relevant to detecting the above world states and situations.

- In step 6, we use the context of the particular task activities to establish the distances and, therefore, the resolutions at which the relevant objects and entities must be measured and recognized by the sensory processing component. This establishes a set of requirements and/or specifications for the sensor system to support each subtask activity.

Real-time Control System Software

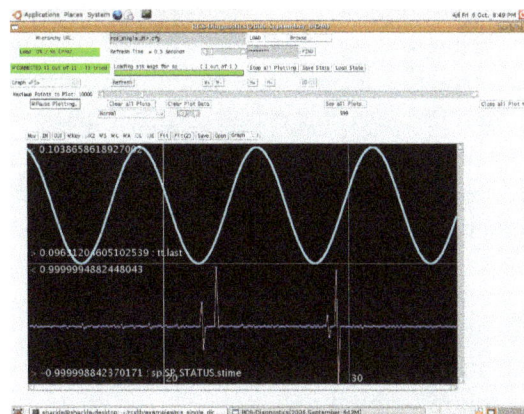

Real-Time Control Systems Software.

Based on the RCS Reference Model Architecture the NIST has developed a Real-time Control System Software Library. This is an archive of free C++, Java and Ada code, scripts, tools, makefiles, and documentation developed to aid programmers of software to be used in real-time control systems, especially those using the Reference Model Architecture for Intelligent Systems Design.

Resilient Control Systems

In our modern society, computerized or digital control systems have been used to reliably automate many of the industrial operations that we take for granted, from the power plant to the automobiles we drive. However, the complexity of these systems and how the designers integrate them, the roles and responsibilities of the humans that interact with the systems, and the cyber security of these highly networked systems has led to a new paradigm in research philosophy for next generation control systems. Resilient Control Systems consider all of these elements and those disciplines that contribute to a more effective design, such as cognitive psychology, computer science, and control engineering to develop interdisciplinary solutions. These solutions consider such things such as how to tailor the control system operating displays to best enable the user to make an accurate and reproducible response, how to design in cyber security protections such that the system defends itself from attack by changing its behaviors, and how to better integrate widely distributed computer control systems to prevent cascading failures that result in disruptions to critical industrial operations. In the context of cyber-physical systems, resilient control systems are an aspect that focuses on the unique interdependencies of a control system, as compared to information technology computer systems and networks, due to its importance in operating our critical industrial operations.

Introduction

Originally intended to provide a more efficient mechanism for controlling industrial operations, the development of digital control systems allowed for flexibility in integrating distributed sensors and operating logic while maintaining a centralized interface for human monitoring and interaction. This ease of readily adding sensors and logic through software, which was once done with relays and isolated analog instruments, has led to wide acceptance and integration of these systems in all industries. However, these digital control systems have often been integrated in phases to cover different aspects of an industrial operation, connected over a network, and leading to a complex interconnected and interdependent system. While the control theory applied is often nothing more than a digital version of their analog counterparts, the dependence of digital control systems upon the communications networks, has precipitated the need for cybersecurity due to potential effects on confidentiality, integrity and availability of the information. To achieve resilience in the next generation of control systems, therefore, addressing the complex control system interdependencies, including the human systems interaction and cyber security, will be a recognized challenge.

Defining Resilience

Research in resilience engineering over the last decade has focused in two areas, organizational and information technology. Organizational resilience considers the ability of an organization to adapt and survive in the face of threats, including the prevention or mitigation of unsafe, hazardous or compromising conditions that threaten its very existence. Information technology resilience has been considered from a number of standpoints . Networking resilience has been considered as quality of service . Computing has considered such issues as dependability and performance in the face of unanticipated changes. However, based upon the application of control dynamics to industrial processes, functionality and determinism are primary considerations that are not captured by the traditional objectives of information technology.

Considering the paradigm of control systems, one definition has been suggested that "Resilient control systems are those that tolerate fluctuations via their structure, design parameters, control structure and control parameters". However, this definition is taken from the perspective of control theory application to a control system. The consideration of the malicious actor and cyber security are not directly considered, which might suggest the definition, "an effective reconstitution of control under attack from intelligent adversaries," which was proposed. However, this definition focuses only on resilience in response to a malicious actor. To consider the cyber-physical aspects of control system, a definition for resilience considers both benign and malicious human interaction, in addition to the complex interdependencies of the control system application .

The use of the term "recovery" has been used in the context of resilience, paralleling the response of a rubber ball to stay intact when a force is exerted on it and recover its original dimensions after the force is removed. Considering the rubber ball in terms of a system, resilience could then be defined as its ability to maintain a desired level of performance or normalcy without irrecoverable consequences. While resilience in this context is based upon the yield strength of the ball, control systems require an interaction with the environment, namely the sensors, valves, pumps that make up the industrial operation. To be reactive to this environment, control systems require an awareness of its state to make corrective changes to the industrial process to maintain normalcy. With this in mind, in consideration of the discussed cyber-physical aspects of human systems integration and cyber security, as well as other definitions for resilience at a broader critical infrastructure level, the following can be deduced as a definition of a resilient control system:

"A resilient control system is one that maintains state awareness and an accepted level of operational normalcy in response to disturbances, including threats of an unexpected and malicious nature"

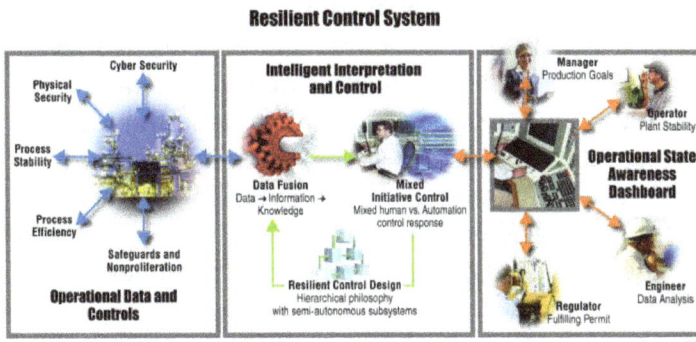

Fig. 1. Resilient Control System Framework.

Considering the flow of a digital control system as a basis, a resilient control system framework can be designed. Referring to the left side of Fig. 1, a resilient control system holistically considers the measures of performance or normalcy for the state space. At the center, an understanding of performance and priority provide the basis for an appropriate response by a combination of human and automation, embedded within a multi-agent, semi-autonomous framework. Finally, to the right, information must be tailored to the consumer to address the need and position a desirable response. Several examples or scenarios of how resilience differs and provides benefit to control system design are available in the literature.

Areas of Resilience

Some primary tenets of resilience, as contrasted to traditional reliability, have presented themselves in considering an integrated approach to resilient control systems. These cyber-physical tenants complement the fundamental concept of dependable or reliable computing by characterizing resilience in regard to control system concerns, including design considerations that provide a level of understanding and assurance in the safe and secure operation of an industrial facility. These tenants are discussed individually below to summarize some of the challenges to address in order to achieve resilience.

Human Systems

The benign human has an ability to quickly understand novel solutions, and provide the ability to adapt to unexpected conditions. This behavior can provide additional resilience to a control system, but reproducibly predicting human behavior is a continuing challenge. The ability to capture historic human preferences can be applied to bayesian inference and bayesian belief networks, but ideally a solution would consider direct understanding of human state using sensors such as an EEG. Considering control system design and interaction, the goal would be to tailor the amount of automation necessary to achieve some level of optimal resilience for this mixed initiative response. Presented to the human would be that actionable information that provides the basis for a targeted, reproducible response.

Cyber Security

In contrast to the challenges of prediction and integration of the benign human with control systems, the abilities of the malicious actor (or hacker) to undermine desired control system behavior also create a significant challenge to control system resilience. Application of dynamic probabilistic risk analysis used in human reliability can provide some basis for the benign actor. However, the decidedly malicious intentions of an adversarial individual, organization or nation make the modeling of the human variable in both objectives and motives. However, in defining a control system response to such intentions, the malicious actor looks forward to some level of recognized behavior to gain an advantage and provide a pathway to undermining the system. Whether performed separately in preparation for a cyber attack, or on the system itself, these behaviors can provide opportunity for a successful attack without detection. Therefore, in considering resilient control system architecture, atypical designs that imbed active and passively implemented randomization of attributes, would be suggested to reduce this advantage.

Complex Networks and Networked Control Systems

While much of the current critical infrastructure is controlled by a web of interconnected control systems, either architecture termed as distributed control systems (DCS) or supervisory control and data acquisition (SCADA), the application of control is moving toward a more decentralized state. In moving to a smart grid, the complex interconnected nature of individual homes, commercial facilities and diverse power generation and storage creates an opportunity and a challenge to ensuring that the resulting system is more resilient to threats. The ability to operate these systems to achieve a global optimum for multiple considerations, such as overall efficiency, stability and security, will require mechanisms to holistically design complex networked control systems. Multi-agent methods suggest a mechanism to tie a global objective to distributed assets, allowing for management and coordination of assets for optimal benefit and semi-autonomous, but constrained controllers that can react rapidly to maintain resilience for rapidly changing conditions.

Base Metrics for Resilient Control Systems

Establishing a metric that can capture the resilience attributes can be complex, at least if considered based upon differences between the interactions or interdependencies. Evaluating the control, cyber and cognitive disturbances, especially if considered from a disciplinary standpoint, leads to measures that already had been established. However, if the metric were instead based upon a normalizing dynamic attribute, such a performance characteristic that can be impacted by degradation, an alternative is suggested. Specifically, applications of base metrics to resilience characteristics are given as follows for type of disturbance:

- Physical Disturbances:
    - Time Latency Affecting Stability
    - Data Integrity Affecting Stability
- Cyber Disturbances:
    - Time Latency
    - Data Confidentiality, Integrity and Availability
- Cognitive Disturbances:
    - Time Latency in Response
    - Data Digression from Desired Response

Such performance characteristics exist with both time and data integrity. Time, both in terms of delay of mission and communications latency, and data, in terms of corruption or modification, are normalizing factors. In general, the idea is to base the metric on "what is expected" and not necessarily the actual initiator to the degradation. Considering time as a metrics basis, resilient and un-resilient systems can be observed in Fig. 2.

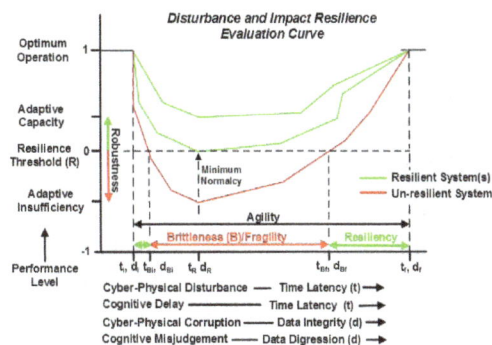

Fig. 2. Resilience Base Metrics.

Dependent upon the abscissa metrics chosen, Fig. 2 reflects a generalization of the resiliency of a system. Several common terms are represented on this graphic, including robustness, agility, adaptive capacity, adaptive insufficiency, resiliency and brittleness. To overview the definitions of these terms, the following explanations of each is provided below:

- Agility: The derivative of the disturbance curve. This average defines the ability of the system to resist degradation on the downward slope, but also to recover on the upward. Primarily considered a time based term that indicates impact to mission.

- Adaptive Capacity: The ability of the system to adapt or transform from impact and maintain minimum normalcy. Considered a value between 0 and 1, where 1 is fully operational and 0 is the resilience threshold.

- Adaptive Insufficiency: The inability of the system to adapt or transform from impact, indicating an unacceptable performance loss due to the disturbance. Considered a value between 0 and -1, where 0 is the resilience threshold and -1 is total loss of operation.

- Brittleness: The area under the disturbance curve as intersected by the resilience threshold. This indicates the impact from the loss of operational normalcy.

- Resiliency: The converse of brittleness, which for a resilience system is "zero" loss of minimum normalcy.

- Robustness: A positive or negative number associated with the area between the disturbance curve and the resilience threshold, indicating either the capacity or insufficiency, respectively.

On the abscissa of Fig. 2, it can be recognized that cyber and cognitive influences can affect both the data and the time, which underscores the relative importance of recognizing these forms of degradation in resilient control designs. For cybersecurity, a single cyberattack can degrade a control system in multiple ways. Additionally, control impacts can be characterized as indicated. While these terms are fundamental and seem of little value for those correlating impact in terms like cost, the development of use cases provide a means by which this relevance can be codified. For example, given the impact to system dynamics or data, the performance of the control loop can be directly ascertained and show approach to instability and operational impact.

Examples of Resilient Control System Developments

1) When considering the current digital control system designs, the cyber security of these systems is dependent upon what is considered border protections, i.e., firewalls, passwords, etc. If a malicious actor compromised the digital control system for an industrial operation by a man-in-the-middle attack, data can be corrupted with the control system. The industrial facility operator would have no way of knowing the data has been compromised, until someone such as a security engineer recognized the attack was occurring. As operators are trained to provide a prompt, appropriate response to stabilize the industrial facility, there is a likelihood that the corrupt data would lead to the operator reacting to the situation and lead to a plant upset. In a resilient control system, as per Fig. 1, cyber and physical data is fused to recognize anomalous situations and warn the operator.

2) As our society becomes more automated for a variety of drivers, including energy efficiency, the need to implement ever more effective control algorithms naturally follow. However, advanced control algorithms are dependent upon data from multiple sensors to predict the behaviors of the industrial operation and make corrective responses. This type of system can become very brittle, insofar as any unrecognized degradation in the sensor itself can lead to incorrect responses by the control algorithm and potentially a worsened condition relative to the desired operation for the industrial facility. Therefore, implementation of advanced control algorithms in a resilient control system also requires the implementation of diagnostic and prognostic architectures to recognize sensor degradation, as well as failures with industrial process equipment associated with the control algorithms.

Resilient Control System Solutions and the Need for Interdisciplinary Education

In our world of advancing automation, our dependence upon these advancing technologies and the skill sets needed to keep the United States at the forefront of innovation. The challenges may appear rooted in design of better means to better control our infrastructures for greater safety and efficiency in generation and use of energy. However, the evolution of the technologies developed to achieve the current design of automation has achieved a complex environment where a cyber-attack, human error in design or operation, or a damaging storm can wreak havoc on the infrastructure we depend as a Nation. The next generation of systems will need to consider the broader picture to ensure as a path forward, failures do not lead to ever greater catastrophic events. As a critical resource are the students of tomorrow who will be expected to advance these designs, and require both a perspective on the challenges and the contributions of others to fulfill the need. Addressing this need, courses have been developed to provide the perspectives and relevant examples to overview the issues and provide opportunity to create resilient solutions at such universities as George Mason University and Northeastern. The tie to critical infrastructure operations is an important aspect of these courses.

Through the development of technologies designed to set the stage for next generation automation, it has become evident that effective teams are comprised several disciplines. However, developing a level of effectiveness can be time consuming, and when done in a professional environment can expend a lot of energy and time that provides little obvious benefit to the desired outcome. It is clear that the earlier these STEM disciplines can be successfully integrated, the more effective they are at recognizing each other's contributions and working together to achieve a common set of goals in the professional world. Team competition at venues such as Resilience Week will be a natural outcome of developing such an environment, allowing interdisciplinary participation and providing an exciting challenge to motivate students to pursue a STEM education.

Standardizing Resilience and Resilient Control System Principles

Standards and policy that define resilience nomenclature and metrics are needed to establish a value proposition for investment, which includes government, academia and industry. The IEEE Industrial Electronics Society has taken the lead in forming a technical committee toward this end. The purpose of this committee will be to establish metrics and standards associated with codifying promising technologies that promote resilience in automation. This effort is distinct from more supply chain community focus on resilience and security, such as the efforts of ISO and NIST

References

- Liptak, Bela (1995). Instrument Engineers' Handbook: Process Control. Radnor, Pennsylvania: Chilton Book Company. pp. 20–29. ISBN 0-8019-8242-1.

- Tan, Kok Kiong; Wang Qing-Guo; Hang Chang Chieh (1999). Advances in PID Control. London, UK: Springer-Verlag. ISBN 1-85233-138-0.

- Dominique Brière, Christian Favre, Pascal Traverse, Electrical Flight Controls, From Airbus A320/330/340 to Future Military Transport Aircraft: A Family of Fault-Tolerant Systems, chapitre 12 du Avionics Handbook, Cary Spitzer ed., CRC Press 2001, ISBN 0-8493-8348-X.

- Whitcomb, Randall L. Cold War Tech War: The Politics of America's Air Defense. Apogee Books, Burlington, Ontario, Canada 2008. Pages 134, 163. ISBN 978-1-894959-77-3

- Ian Moir; Allan G. Seabridge; Malcolm Jukes (2003). Civil Avionics Systems. London (iMechE): Professional Engineering Publishing Ltd. ISBN 1-86058-342-3.

- "Air France 447 Flight-Data Recorder Transcript – What Really Happened Aboard Air France 447". Popular Mechanics. 6 December 2011. Retrieved 7 July 2012.

Tools and Methods of Control Systems

Tools and methods are an important component of any field of study. The following chapter elucidates the various tools and methods of control systems. Some of the tools, like a transducer, are explained in this chapter. The following text also explains to the reader the relevance of control systems.

Open-loop Controller

An open-loop controller, also called a non-feedback controller, is a type of controller that computes its input into a system using only the current state and its model of the system.

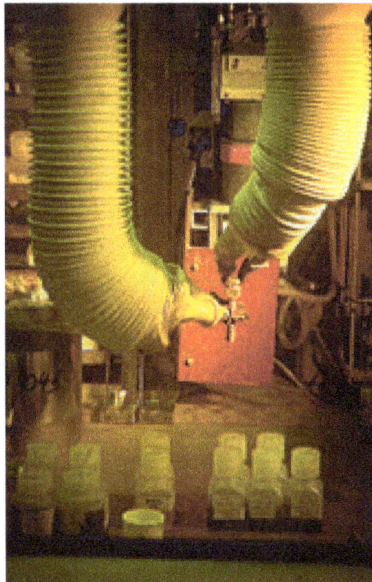

Remote manipulator arms for working with radioactive materials – an open-loop
mechanism, controlled by hand controls

A characteristic of the open-loop controller is that it does not use feedback to determine if its output has achieved the desired goal of the input. This means that the system does not observe the output of the processes that it is controlling. Consequently, a true open-loop system can not engage in machine learning and also cannot correct any errors that it could make. It also may not compensate for disturbances in the system.

Examples

An open-loop controller is often used in simple processes because of its simplicity and low cost, especially in systems where feedback is not critical. A typical example would be a conventional washing machine, for which the length of machine wash time is entirely dependent on the judgment and estimation of the human operator. Generally, to obtain a more accurate or more adaptive control, it is necessary to feed the output of the system back to the inputs of the controller. This type of system is called a closed-loop system.

For example, an irrigation sprinkler system, programmed to turn on at set times could be an example of an open-loop system if it does not measure soil moisture as a form of feedback. Even if rain is pouring down on the lawn, the sprinkler system would activate on schedule, wasting water.

Stepper motors are often used for open-loop control of position. A stepper motor rotates to one of a number of fixed positions, according to its internal construction. Sending a stream of electrical pulses to it causes it to rotate by exactly that many steps, hence the name. Such motors are often used, together with a simple initial datum sensor (a switch that is activated at the machine's home position), for the control of simple robotic machines or even the commonplace inkjet printer head. The drawback of open-loop control of steppers is that if the machine load is too high, or the motor attempts to move too quickly, then steps may be skipped. The controller has no means of detecting this and so the machine continues to run slightly out of adjustment, until reset. For this reason, more complex robots and machine tools instead use servomotors rather than stepper motors, which incorporate encoders and closed-loop controllers.

Open-loop control is useful for well-defined systems where the relationship between input and the resultant state can be modeled by a mathematical formula. For example, determining the voltage to be fed to an electric motor that drives a constant load, in order to achieve a desired speed would be a good application of open-loop control. If the load were not predictable, on the other hand, the motor's speed might vary as a function of the load as well as of the voltage, and an open-loop controller would therefore be insufficient to ensure repeatable control of the velocity.

An example of this is a conveyor system that is required to travel at a constant speed. For a constant voltage, the conveyor will move at a different speed depending on the load on the motor (represented here by the weight of objects on the conveyor). In order for the conveyor to run at a constant speed, the voltage of the motor must be adjusted depending on the load. In this case, a closed-loop control system would be necessary.

Transducer

A transducer is a device that converts one form of energy to another. Usually a transducer converts a signal in one form of energy to a signal in another.

Transducers are often employed at the boundaries of automation, measurement, and control systems, where electrical signals are converted to and from other physical quantities (energy, force, torque, light, motion, position, etc.). The process of converting one form of energy to another is known as transduction.

Transducer Types

Active

Active sensors require an external power sources to operate, which is called an excitation signal. The signal is modulated by the sensor to produce the output signal. For example, a thermistor does not generate any electric signal, but by passing electric current through it, its resistance can be measured by detecting variations in current and/or voltage across the thermistor.

Passive

Passive sensors generate electric signals in response to an external stimulus without the need of an additional energy source. Such examples are a thermocouple, photodiode, and a piezoelectric sensor.

Sensors

A sensor is a device that receives and responds to a signal or stimulus. Transducer is the other term that is sometimes interchangeably used instead of the term sensor, although there are subtle differences. A transducer is a term that can be used for the definition of many devices such as sensors, actuators, or transistors.

Actuators

An actuator is a device that is responsible for moving or controlling a mechanism or system. It is operated by a source of energy, which can be mechanical force, electrical current, hydraulic fluid pressure, or pneumatic pressure, and converts that energy into motion. An actuator is the mechanism by which a control system acts upon an environment. The control system can be simple (a fixed mechanical or electronic system), software-based (e.g. a printer driver, robot control system), a human, or any other input.

Bidirectional

Bidirectional transducers convert physical phenomena to electrical signals and also convert electrical signals into physical phenomena. Examples of inherently bidirectional transducers are antennae, which can convert conducted electrical signals to or from propagating electromagnetic waves, and voice coils, which convert electrical signals into sound (when used in a loudspeaker) or sound into electrical signals (when used in a microphone). Likewise, DC electric motors may be used to generate electrical power if the motor shaft is turned by an external torque.

Ideal Characteristics

- High dynamic range
- High Repeatability
- Low noise
- Low hysteresis

Applications

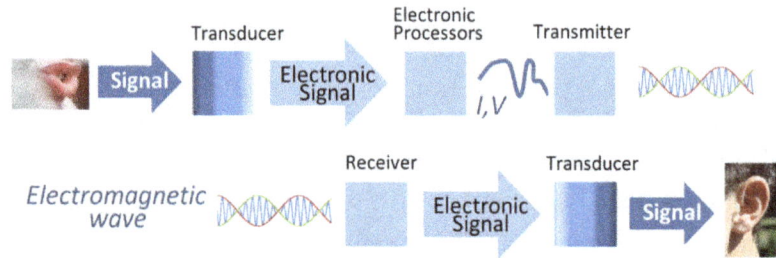

Transducers are used in electronic communications systems to convert signals of various physical forms to electronic signals, and *vice versa*. In this example, the first transducer could be a microphone, and the second transducer could be a speaker.

- Electromagnetic:

  o Antennae – converts propagating electromagnetic waves to and from conducted electrical signals

  o magnetic cartridges – converts relative physical motion to and from electrical signals

  o Tape head, disk read-and-write heads – converts magnetic fields on a magnetic medium to and from electrical signals

  o Hall effect sensors – converts a magnetic field level into an electrical signal

- Electrochemical:

  o pH probes

  o Electro-galvanic fuel cells

  o Hydrogen sensors

- Electromechanical (electromechanical output devices are generically called actuators):

  o Accelerometers

  o Air flow sensors

  o Electroactive polymers

  o Rotary motors, linear motors

  o Galvonometers

  o Linear variable differential transformers or rotary variably differential transformers

  o Load cells – converts force to mV/V electrical signal using strain gauges

  o Microelectromechanical systems

  o Potentiometers (when used for measuring position)

- o   Pressure sensors

- o   String potentiometers

- o   Tactile sensors

- o   Vibration powered generators

- Electroacoustic:

  - o   Loudspeakers, earphones – converts electrical signals into sound (amplified signal → magnetic field → motion → air pressure)

  - o   Microphones – converts sound into an electrical signal (air pressure → motion of conductor/coil → magnetic field → electrical signal)

  - o   Pickup (music technology) – converts motion of metal strings into an electrical signal (magnetism → electrical signal)

  - o   Tactile transducers – converts electrical signal into vibration (electrical signal → vibration)

  - o   Piezoelectric crystals – converts deformations of solid-state crystals (vibrations) to and from electrical signals

  - o   Geophones – converts a ground movement (displacement) into voltage (vibrations → motion of conductor/coil → magnetic field → signal)

  - o   Gramophone pickups – (air pressure → motion → magnetic field → electrical signal)

  - o   Hydrophones – converts changes in water pressure into an electrical signal

  - o   Sonar transponders (water pressure → motion of conductor/coil → magnetic field → electrical signal)

  - o   Ultrasonic transceivers, transmitting ultrasound (transduced from electricity) as well as receiving it after sound reflection from target objects, availing for imaging of those objects.

- Electro-optical (Photoelectric):

  - o   Fluorescent lamps – converts electrical power into incoherent light

  - o   Incandescent lamps – converts electrical power into incoherent light

  - o   Light-emitting diodes – converts electrical power into incoherent light

  - o   Laser diodes – converts electrical power into coherent light

  - o   Photodiodes, photoresistors, phototransistors, photomultipliers – converts changing light levels into electrical signals

  - o   Photodetector or photoresistor or light dependent resistor (LDR) – converts changes in light levels into changes in electrical resistance

- o   Cathode-ray tubes (CRT) – converts electrical signals into visual signals

- Electrostatic:

  - o   Electrometers

- Thermoelectric:

  - o   Resistance temperature detectors (RTD) – converts temperature into an electrical resistance signal

  - o   Thermocouples – converts relative temperatures of metallic junctions to electrical voltage

  - o   Thermistors (includes PTC resistor and NTC resistor)

- Radioacoustic:

  - o   Geiger-Müller tubes – converts incident ionizing radiation to an electrical impulse signal

  - o   Radio receivers converts electromagnetic transmissions to electrical signals.

  - o   Radio transmitters converts electrical signals to electromagnetic transmissions.

Adaptive Control

Adaptive control is the control method used by a controller which must adapt to a controlled system with parameters which vary, or are initially uncertain. For example, as an aircraft flies, its mass will slowly decrease as a result of fuel consumption; a control law is needed that adapts itself to such changing conditions. Adaptive control is different from robust control in that it does not need *a priori* information about the bounds on these uncertain or time-varying parameters; robust control guarantees that if the changes are within given bounds the control law need not be changed, while adaptive control is concerned with control law changing themselves.

Parameter Estimation

The foundation of adaptive control is parameter estimation. Common methods of estimation include recursive least squares and gradient descent. Both of these methods provide update laws which are used to modify estimates in real time (i.e., as the system operates). Lyapunov stability is used to derive these update laws and show convergence criterion (typically persistent excitation). Projection (mathematics) and normalization are commonly used to improve the robustness of estimation algorithms. It is also called adjustable control.

Classification of Adaptive Control Techniques

In general one should distinguish between:

1. Feedforward Adaptive Control

2. Feedback Adaptive Control

as well as between

1. Direct Methods and

2. Indirect Methods

Direct methods are ones wherein the estimated parameters are those directly used in the adaptive controller. In contrast, indirect methods are those in which the estimated parameters are used to calculate required controller parameters

There are several broad categories of feedback adaptive control (classification can vary):

- Dual Adaptive Controllers [based on Dual control theory]

    o Optimal Dual Controllers [difficult to design]

    o Suboptimal Dual Controllers

- Nondual Adaptive Controllers

    o Adaptive Pole Placement

    o Extremum Seeking Controllers

    o Iterative learning control

    o Gain scheduling

    o Model Reference Adaptive Controllers (MRACs) [incorporate a *reference model* defining desired closed loop performance]

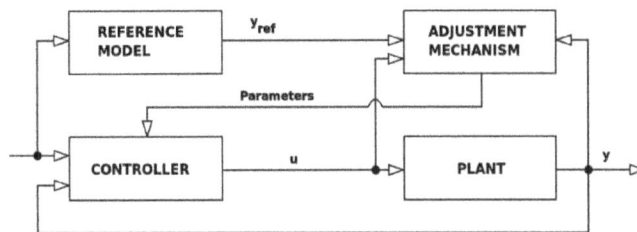

MODEL REFERENCE ADAPTIVE CONTROL (MRAC)

MRAC

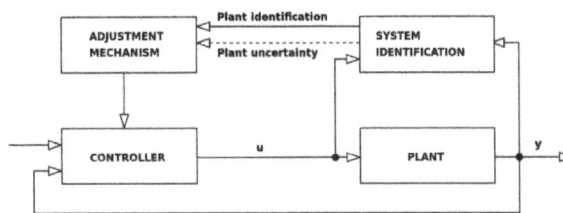

MODEL IDENTIFICATION ADAPTIVE CONTROL (MIAC)

MIAC

- Gradient Optimization MRACs [use local rule for adjusting params when performance differs from reference. Ex.: "MIT rule".]

- Stability Optimized MRACs

o Model Identification Adaptive Controllers (MIACs) [perform System identification while the system is running]

- Cautious Adaptive Controllers [use current SI to modify control law, allowing for SI uncertainty]

- Certainty Equivalent Adaptive Controllers [take current SI to be the true system, assume no uncertainty]

  - Nonparametric Adaptive Controllers

  - Parametric Adaptive Controllers

    - Explicit Parameter Adaptive Controllers

    - Implicit Parameter Adaptive Controllers

o Multiple Models [Use large number of models, which are distributed in the region of uncertainty, and based on the responses of the plant and the models. One model is chosen at every instant, which is closest to the plant according to some metric.]

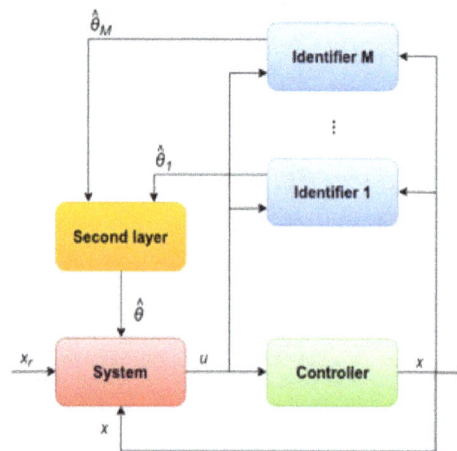

Adaptive Control with Multiple Models

Some special topics in adaptive control can be introduced as well:

1. Adaptive Control Based on Discrete-Time Process Identification

2. Adaptive Control Based on the Model Reference Technique

3. Adaptive Control based on Continuous-Time Process Models

4. Adaptive Control of Multivariable Processes

5. Adaptive Control of Nonlinear Processes

Adaptive control has even been merged with intelligent techniques such as fuzzy and neural networks and the new terms like fuzzy adaptive control has been generated.

Applications

When designing adaptive control systems, special consideration is necessary of convergence and robustness issues. Lyapunov stability is typically used to derive control adaptation laws and show convergence.

Typical applications of adaptive control are (in general):

- Self-tuning of subsequently fixed linear controllers during the implementation phase for one operating point;

- Self-tuning of subsequently fixed robust controllers during the implementation phase for whole range of operating points;

- Self-tuning of fixed controllers on request if the process behaviour changes due to ageing, drift, wear etc.;

- Adaptive control of linear controllers for nonlinear or time-varying processes;

- Adaptive control or self-tuning control of nonlinear controllers for nonlinear processes;

- Adaptive control or self-tuning control of multivariable controllers for multivariable processes (MIMO systems);

Usually these methods adapt the controllers to both the process statics and dynamics. In special cases the adaptation can be limited to the static behavior alone, leading to adaptive control based on characteristic curves for the steady-states or to extremum value control, optimizing the steady state. Hence, there are several ways to apply adaptive control algorithms.

SCADA

Supervisory control and data acquisition (SCADA) is a system for remote monitoring and control that operates with coded signals over communication channels (using typically one communication channel per remote station).

The control system may be combined with a data acquisition system by adding the use of coded signals over communication channels to acquire information about the status of the remote equipment for display or for recording functions. It is a type of industrial control system (ICS). Industrial control systems are computer-based systems that monitor and control industrial processes that exist in the physical world. SCADA systems historically distinguish themselves from other ICS systems by being large-scale processes that can include multiple sites, and large distances. These processes include industrial, infrastructure, and facility-based processes, as described below:

- Industrial processes include those of manufacturing, production, power generation, fabrication, and refining, and may run in continuous, batch, repetitive, or discrete modes.

- Infrastructure processes may be public or private, and include water treatment and distribution, wastewater collection and treatment, oil and gas pipelines, electrical power transmission and distribution, wind farms, civil defense siren systems, and large communication systems.

- Facility processes occur both in public facilities and private ones, including buildings, airports, ships, and space stations. They monitor and control heating, ventilation, and air conditioning systems (HVAC), access, and energy consumption.

Common System Components

A SCADA system usually consists of the following subsystems:

- Remote terminal units (RTUs) connect to sensors in the process and convert sensor signals to digital data. They have telemetry hardware capable of sending digital data to the supervisory system, as well as receiving digital commands from the supervisory system. RTUs often have embedded control capabilities such as ladder logic in order to accomplish boolean logic operations.

- Programmable logic controller (PLCs) connect to sensors in the process and convert sensor signals to digital data. PLCs have more sophisticated embedded control capabilities (typically one or more IEC 61131-3 programming languages) than RTUs. PLCs do not have telemetry hardware, although this functionality is typically installed alongside them. PLCs are sometimes used in place of RTUs as field devices because they are more economical, versatile, flexible, and configurable.

- A telemetry system is typically used to connect PLCs and RTUs with control centers, data warehouses, and the enterprise. Examples of wired telemetry media used in SCADA systems include leased telephone lines and WAN circuits. Examples of wireless telemetry media used in SCADA systems include satellite (VSAT), licensed and unlicensed radio, cellular and microwave.

- A data acquisition server is a software service which uses industrial protocols to connect software services, via telemetry, with field devices such as RTUs and PLCs. It allows clients to access data from these field devices using standard protocols.

- A human–machine interface or HMI is the apparatus or device which presents processed data to a human operator, and through this, the human operator monitors and interacts with the process. The HMI is a client that requests data from a data acquisition server or in most installations the HMI is the graphical user interface for the operator, collects all data from external devices, creates reports, performs alarming, sends notifications, etc.

- A historian is a software service which accumulates time-stamped data, boolean events, and boolean alarms in a database which can be queried or used to populate graphic trends in the HMI. The historian is a client that requests data from a data acquisition server.

- A supervisory (computer) system, gathering (acquiring) data on the process and sending commands (control) to the SCADA system.

- Communication infrastructure connecting the supervisory system to the remote terminal units.

- Various processes and analytical instrumentation.

Systems Concepts

The term SCADA usually refers to centralized systems which monitor and control entire sites, or complexes of systems spread out over large areas (anything from an industrial plant to a nation). Most control actions are performed automatically by RTUs or by PLCs. Host control functions are usually restricted to basic overriding or *supervisory* level intervention. For example, a PLC may control the flow of cooling water through part of an industrial process, but the SCADA system may allow operators to change the set points for the flow, and enable alarm conditions, such as loss of flow and high temperature, to be displayed and recorded. The feedback control loop passes through the RTU or PLC, while the SCADA system monitors the overall performance of the loop.

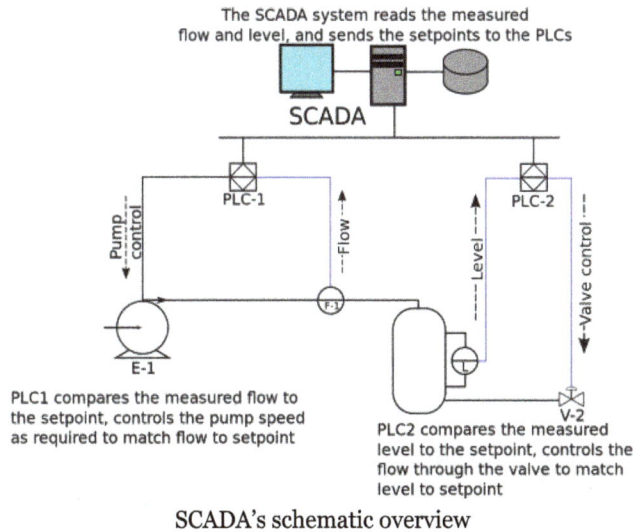

SCADA's schematic overview

Data acquisition begins at the RTU or PLC level and includes meter readings and equipment status reports that are communicated to SCADA as required. Data is then compiled and formatted in such a way that a control room operator using the HMI can make supervisory decisions to adjust or override normal RTU (PLC) controls. Data may also be fed to a Historian, often built on a commodity Database Management System, to allow trending and other analytical auditing.

SCADA systems typically implement a distributed database, commonly referred to as a *tag database*, which contains data elements called *tags* or *points*. A point represents a single input or output value monitored or controlled by the system. Points can be either "hard" or "soft". A hard point represents an actual input or output within the system, while a soft point results from logic and math operations applied to other points. (Most implementations conceptually remove the distinction by making every property a "soft" point expression, which may, in the simplest case, equal a single hard point.) Points are normally stored as value-timestamp pairs: a value, and the timestamp when it was recorded or calculated. A series of value-timestamp pairs gives the history of that point. It is also common to store additional metadata with tags, such as the path to a field device or PLC register, design time comments, and alarm information.

SCADA systems are significantly important systems used in national infrastructures such as electric grids, water supplies and pipelines. However, SCADA systems may have security vulnerabilities, so the systems should be evaluated to identify risks and solutions implemented to mitigate those risks.

Human–machine Interface

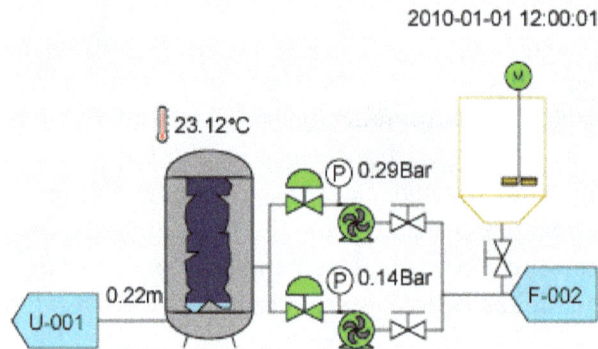

Typical basic SCADA animations

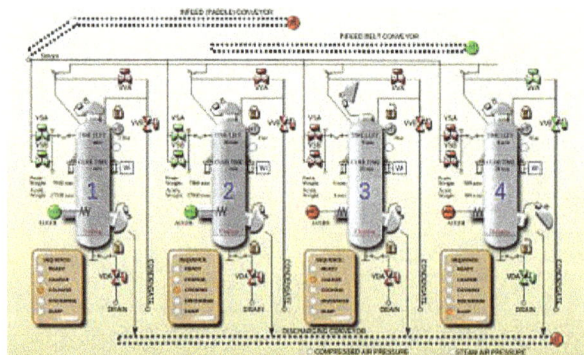

More complex SCADA animation

A human–machine interface (HMI) is the input-output device through which the human operator controls the process, and which presents process data to a human operator.

HMI (human machine interface) is usually linked to the SCADA system's databases and software programs, to provide trending, diagnostic data, and management information such as scheduled maintenance procedures, logistic information, detailed schematics for a particular sensor or machine, and expert-system troubleshooting guides.

The HMI system usually presents the information to the operating personnel graphically, in the form of a mimic diagram. This means that the operator can see a schematic representation of the plant being controlled. For example, a picture of a pump connected to a pipe can show the operator that the pump is running and how much fluid it is pumping through the pipe at the moment. The operator can then switch the pump off. The HMI software will show the flow rate of the fluid in the pipe decrease in real time. Mimic diagrams may consist of line graphics and schematic symbols to represent process elements, or may consist of digital photographs of the process equipment overlain with animated symbols.

The HMI package for the SCADA system typically includes a drawing program that the operators or system maintenance personnel use to change the way these points are represented in the interface. These representations can be as simple as an on-screen traffic light, which represents the state of an actual traffic light in the field, or as complex as a multi-projector display representing the position of all of the elevators in a skyscraper or all of the trains on a railway.

An important part of most SCADA implementations is alarm handling. The system monitors whether certain alarm conditions are satisfied, to determine when an alarm event has occurred. Once an alarm event has been detected, one or more actions are taken (such as the activation of one or more alarm indicators, and perhaps the generation of email or text messages so that management or remote SCADA operators are informed). In many cases, a SCADA operator may have to acknowledge the alarm event; this may deactivate some alarm indicators, whereas other indicators remain active until the alarm conditions are cleared. Alarm conditions can be explicit—for example, an alarm point is a digital status point that has either the value NORMAL or ALARM that is calculated by a formula based on the values in other analogue and digital points—or implicit: the SCADA system might automatically monitor whether the value in an analogue point lies outside high and low- limit values associated with that point. Examples of alarm indicators include a siren, a pop-up box on a screen, or a coloured or flashing area on a screen (that might act in a similar way to the "fuel tank empty" light in a car); in each case, the role of the alarm indicator is to draw the operator's attention to the part of the system 'in alarm' so that appropriate action can be taken. In designing SCADA systems, care must be taken when a cascade of alarm events occurs in a short time, otherwise the underlying cause (which might not be the earliest event detected) may get lost in the noise. Unfortunately, when used as a noun, the word 'alarm' is used rather loosely in the industry; thus, depending on context it might mean an alarm point, an alarm indicator, or an alarm event.

Hardware Solutions

SCADA solutions often have distributed control system (DCS) components. Use of "smart" RTUs or PLCs, which are capable of autonomously executing simple logic processes without involving the master computer, is increasing. A standardized control programming language, IEC 61131-3 (a suite of 5 programming languages including function block, ladder, structured text, sequence function charts and instruction list), is frequently used to create programs which run on these RTUs and PLCs. Unlike a procedural language such as the C programming language or FORTRAN, IEC 61131-3 has minimal training requirements by virtue of resembling historic physical control arrays. This allows SCADA system engineers to perform both the design and implementation of a program to be executed on an RTU or PLC. A programmable automation controller (PAC) is a compact controller that combines the features and capabilities of a PC-based control system with that of a typical PLC. PACs are deployed in SCADA systems to provide RTU and PLC functions. In many electrical substation SCADA applications, "distributed RTUs" use information processors or station computers to communicate with digital protective relays, PACs, and other devices for I/O, and communicate with the SCADA master in lieu of a traditional RTU.

Since about 1998, virtually all major PLC manufacturers have offered integrated HMI/SCADA systems, many of them using open and non-proprietary communications protocols. Numerous specialized third-party HMI/SCADA packages, offering built-in compatibility with most major

PLCs, have also entered the market, allowing mechanical engineers, electrical engineers and technicians to configure HMIs themselves, without the need for a custom-made program written by a software programmer. The Remote Terminal Unit (RTU) connects to physical equipment. Typically, an RTU converts the electrical signals from the equipment to digital values such as the open/closed status from a switch or a valve, or measurements such as pressure, flow, voltage or current. By converting and sending these electrical signals out to equipment the RTU can control equipment, such as opening or closing a switch or a valve, or setting the speed of a pump.

Supervisory Station

The term *supervisory station* refers to the servers and software responsible for communicating with the field equipment (RTUs, PLCs, SENSORS etc.), and then to the HMI software running on workstations in the control room, or elsewhere. In smaller SCADA systems, the master station may be composed of a single PC. In larger SCADA systems, the master station may include multiple servers, distributed software applications, and disaster recovery sites. To increase the integrity of the system the multiple servers will often be configured in a dual-redundant or hot-standby formation providing continuous control and monitoring in the event of a server malfunction or breakdown.

Operational Philosophy

For some installations, the costs that would result from the control system failing are extremely high. Hardware for some SCADA systems is ruggedized to withstand temperature, vibration, and voltage extremes. In the most critical installations, reliability is enhanced by having redundant hardware and communications channels, up to the point of having multiple fully equipped control centres. A failing part can be quickly identified and its functionality automatically taken over by backup hardware. A failed part can often be replaced without interrupting the process. The reliability of such systems can be calculated statistically and is stated as the mean time to failure, which is a variant of Mean Time Between Failures (MTBF). The calculated mean time to failure of such high reliability systems can be on the order of centuries.

Communication Infrastructure and Methods

SCADA systems have traditionally used combinations of radio and direct wired connections, although SONET/SDH is also frequently used for large systems such as railways and power stations. The remote management or monitoring function of a SCADA system is often referred to as telemetry. Some users want SCADA data to travel over their pre-established corporate networks or to share the network with other applications. The legacy of the early low-bandwidth protocols remains, though.

SCADA protocols are designed to be very compact. Many are designed to send information only when the master station polls the RTU. Typical legacy SCADA protocols include Modbus RTU, RP-570, Profibus and Conitel. These communication protocols are all SCADA-vendor specific but are widely adopted and used. Standard protocols are IEC 60870-5-101 or 104, IEC 61850 and DNP3. These communication protocols are standardized and recognized by all major SCADA vendors. Many of these protocols now contain extensions to operate over TCP/IP. Although the use of conventional networking specifications, such as TCP/IP, blurs the line between traditional and industrial networking, they each fulfill fundamentally differing requirements.

With increasing security demands (such as North American Electric Reliability Corporation (NERC) and Critical Infrastructure Protection (CIP) in the US), there is increasing use of satellite-based communication. This has the key advantages that the infrastructure can be self-contained (not using circuits from the public telephone system), can have built-in encryption, and can be engineered to the availability and reliability required by the SCADA system operator. Earlier experiences using consumer-grade VSAT were poor. Modern carrier-class systems provide the quality of service required for SCADA.

RTUs and other automatic controller devices were developed before the advent of industry wide standards for interoperability. The result is that developers and their management created a multitude of control protocols. Among the larger vendors, there was also the incentive to create their own protocol to "lock in" their customer base. A list of automation protocols is compiled here.

Recently, OLE for process control (OPC) has become a widely accepted solution for intercommunicating different hardware and software, allowing communication even between devices originally not intended to be part of an industrial network.

Scada Architectures

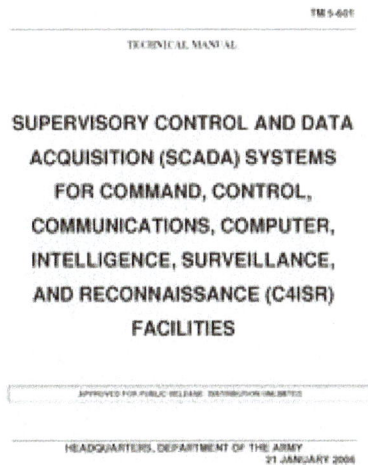

The United States Army's Training Manual 5-601 covers "SCADA Systems for C4ISR Facilities".

SCADA systems have evolved through four generations as follows:

First Generation: "Monolithic"

Early SCADA system computing was done by large minicomputers. Common network services did not exist at the time SCADA was developed. Thus SCADA systems were independent systems with no connectivity to other systems. The communication protocols used were strictly proprietary at that time. The first-generation SCADA system redundancy was achieved using a back-up mainframe system connected to all the Remote Terminal Unit sites and was used in the event of failure of the primary mainframe system. Some first generation SCADA systems were developed as "turn key" operations that ran on minicomputers such as the PDP-11 series made by the Digital Equipment Corporation..

Second Generation: "Distributed"

SCADA information and command processing was distributed across multiple stations which were connected through a LAN. Information was shared in near real time. Each station was responsible for a particular task, which reduced the cost as compared to First Generation SCADA. The network protocols used were still not standardized. Since these protocols were proprietary, very few people beyond the developers knew enough to determine how secure a SCADA installation was. Security of the SCADA installation was usually overlooked.

Third Generation: "Networked"

Similar to a distributed architecture, any complex SCADA can be reduced to simplest components and connected through communication protocols. In the case of a networked design, the system may be spread across more than one LAN network called a process control network (PCN) and separated geographically. Several distributed architecture SCADAs running in parallel, with a single supervisor and historian, could be considered a network architecture. This allows for a more cost effective solution in very large scale systems.

Fourth Generation: "Internet of Things"

With the commercial availability of cloud computing, SCADA systems have increasingly adopted Internet of Things technology to significantly reduce infrastructure costs and increase ease of maintenance and integration. As a result, SCADA systems can now report state in near real-time and use the horizontal scale available in cloud environments to implement more complex control algorithms than are practically feasible to implement on traditional programmable logic controllers. Further, the use of open network protocols such as TLS inherent in the Internet of Things technology, provides a more readily comprehensible and manageable security boundary than the heterogeneous mix of proprietary network protocols typical of many decentralized SCADA implementations. One such example of this technology is an innovative approach to rainwater harvesting through the implementation of real time controls (RTC).

This decentralization of data also requires a different approach to SCADA than traditional PLC based programs. When a SCADA system is used locally, the preferred methodology involves binding the graphics on the user interface to the data stored in specific PLC memory addresses. However, when the data comes from a disparate mix of sensors, controllers and databases (which may be local or at varied connected locations), the typical 1 to 1 mapping becomes problematic. A solution to this is data modeling, a concept derived from object oriented programming.

In a data model, a virtual representation of each device is constructed in the SCADA software. These virtual representations ("models") can contain not just the address mapping of the device represented, but also any other pertinent information (web based info, database entries, media files, etc.) that may be used by other facets of the SCADA/IoT implementation. As the increased complexity of the Internet of Things renders traditional SCADA increasingly "house-bound," and as communication protocols evolve to favor platform-independent, service-oriented architecture (such as OPC UA), it is likely that more SCADA software developers will implement some form of data modeling.

Security Issues

SCADA systems that tie together decentralized facilities such as power, oil, gas pipelines, water distribution and wastewater collection systems were designed to be open, robust, and easily operated and repaired, but not necessarily secure. The move from proprietary technologies to more standardized and open solutions together with the increased number of connections between SCADA systems, office networks and the Internet has made them more vulnerable to types of network attacks that are relatively common in computer security. For example, United States Computer Emergency Readiness Team (US-CERT) released a vulnerability advisory that allowed unauthenticated users to download sensitive configuration information including password hashes on an Inductive Automation Ignition system utilizing a standard attack type leveraging access to the Tomcat Embedded Web server. Security researcher Jerry Brown submitted a similar advisory regarding a buffer overflow vulnerability in a Wonderware InBatchClient ActiveX control. Both vendors made updates available prior to public vulnerability release. Mitigation recommendations were standard patching practices and requiring VPN access for secure connectivity. Consequently, the security of some SCADA-based systems has come into question as they are seen as potentially vulnerable to cyber attacks.

In particular, security researchers are concerned about:

- the lack of concern about security and authentication in the design, deployment and operation of some existing SCADA networks

- the belief that SCADA systems have the benefit of security through obscurity through the use of specialized protocols and proprietary interfaces

- the belief that SCADA networks are secure because they are physically secured

- the belief that SCADA networks are secure because they are disconnected from the Internet.

SCADA systems are used to control and monitor physical processes, examples of which are transmission of electricity, transportation of gas and oil in pipelines, water distribution, traffic lights, and other systems used as the basis of modern society. The security of these SCADA systems is important because compromise or destruction of these systems would impact multiple areas of society far removed from the original compromise. For example, a blackout caused by a compromised electrical SCADA system would cause financial losses to all the customers that received electricity from that source. How security will affect legacy SCADA and new deployments remains to be seen.

There are many threat vectors to a modern SCADA system. One is the threat of unauthorized access to the control software, whether it be human access or changes induced intentionally or accidentally by virus infections and other software threats residing on the control host machine. Another is the threat of packet access to the network segments hosting SCADA devices. In many cases, the control protocol lacks any form of cryptographic security, allowing an attacker to control a SCADA device by sending commands over a network. In many cases SCADA users have assumed that having a VPN offered sufficient protection, unaware that security can be trivially bypassed with physical access to SCADA-related network jacks and switches. Industrial control vendors suggest approaching SCADA security like Information Security with a defense in depth strategy that leverages common IT practices.

The reliable function of SCADA systems in our modern infrastructure may be crucial to public health and safety. As such, attacks on these systems may directly or indirectly threaten public health and safety. Such an attack has already occurred, carried out on Maroochy Shire Council's sewage control system in Queensland, Australia. Shortly after a contractor installed a SCADA system in January 2000, system components began to function erratically. Pumps did not run when needed and alarms were not reported. More critically, sewage flooded a nearby park and contaminated an open surface-water drainage ditch and flowed 500 meters to a tidal canal. The SCADA system was directing sewage valves to open when the design protocol should have kept them closed. Initially this was believed to be a system bug. Monitoring of the system logs revealed the malfunctions were the result of cyber attacks. Investigators reported 46 separate instances of malicious outside interference before the culprit was identified. The attacks were made by a disgruntled ex-employee of the company that had installed the SCADA system. The ex-employee was hoping to be hired by the utility full-time to maintain the system.

In April 2008, the Commission to Assess the Threat to the United States from Electromagnetic Pulse (EMP) Attack issued a Critical Infrastructures Report which discussed the extreme vulnerability of SCADA systems to an electromagnetic pulse (EMP) event. After testing and analysis, the Commission concluded: "SCADA systems are vulnerable to an EMP event. The large numbers and widespread reliance on such systems by all of the Nation's critical infrastructures represent a systemic threat to their continued operation following an EMP event. Additionally, the necessity to reboot, repair, or replace large numbers of geographically widely dispersed systems will considerably impede the Nation's recovery from such an assault."

Many vendors of SCADA and control products have begun to address the risks posed by unauthorized access by developing lines of specialized industrial firewall and VPN solutions for TCP/IP-based SCADA networks as well as external SCADA monitoring and recording equipment. The International Society of Automation (ISA) started formalizing SCADA security requirements in 2007 with a working group, WG4. WG4 "deals specifically with unique technical requirements, measurements, and other features required to evaluate and assure security resilience and performance of industrial automation and control systems devices".

The increased interest in SCADA vulnerabilities has resulted in vulnerability researchers discovering vulnerabilities in commercial SCADA software and more general offensive SCADA techniques presented to the general security community. In electric and gas utility SCADA systems, the vulnerability of the large installed base of wired and wireless serial communications links is addressed in some cases by applying bump-in-the-wire devices that employ authentication and Advanced Encryption Standard encryption rather than replacing all existing nodes.

In June 2010, anti-virus security company VirusBlokAda reported the first detection of malware that attacks SCADA systems (Siemens' WinCC/PCS 7 systems) running on Windows operating systems. The malware is called Stuxnet and uses four zero-day attacks to install a rootkit which in turn logs into the SCADA's database and steals design and control files. The malware is also capable of changing the control system and hiding those changes. The malware was found on 14 systems, the majority of which were located in Iran.

In October 2013 National Geographic released a docudrama titled, "American Blackout" which dealt with a large-scale cyber attack on SCADA and the United States' electrical grid.

Scada In the Workplace

SCADA is one of many tools that can be used while working in an environment where operational duties need to be monitored through electronic communication instead of locally. For example, an operator can position a valve to open or close through SCADA without leaving the control station or the computer. The SCADA system also can switch a pump or motor on or off and has the capability of putting motors on a "Hand" operating status, Off, or Automatic. "Hand" refers to operating the equipment locally, while Automatic has the equipment operate according to set points the operator provides on a computer that can communicate with the equipment through SCADA.

EPICS

The Experimental Physics and Industrial Control System (EPICS) is a software environment used to develop and implement distributed control systems to operate devices such as particle accelerators, telescopes and other large experiments. EPICS also provides SCADA capabilities. The tool is designed to help develop systems which often feature large numbers of networked computers providing control and feedback.

EPICS uses client/server and publish/subscribe techniques to communicate between the various computers. One set of computers (the servers or input/output controllers), collect experiment and control data in real-time using the measurement instruments attached to it. This information is given to another set of computers (the clients) using the Channel Access (CA) network protocol. CA is a high bandwidth networking protocol, which is well suited to soft real-time applications such as scientific experiments.

Look and Feel

EPICS interfaces to the real world with IOCs (Input Output Controllers) . These are either stock-standard PCs, VME, or MicroTCA standard embedded system processors that manage a variety of "plug and play" modules (GPIB, RS-232, IP Carrier etc.) which interface to control system instruments (oscilloscopes, network analyzers) and devices (motors, thermocouples, switches, etc.). Some instruments also can come with EPICS already embedded within them, like certain Oscilloscopes . The IOC holds and runs a database of 'records' which represent either devices or aspects of the devices to be controlled. IOC software used for hard-real-time normally use RTEMS or VxWorks, though work has been ongoing in porting to other systems. Soft real-time IOC software sometimes runs on Linux or Microsoft Windows based machines.

Other computers on the network can interact with the IOC via the concept of channels. For example, take a particle accelerator with shutters between sectors. There would typically be several channels corresponding to a shutter: an output channel to activate shutter motion; an input channel to see the status of the shutter (e.g. shut, open, moving, etc.); and, probably some additional analog input channels representing temperatures and pressures on each side of the shutter. Channel names are typically in the form EQUIPMENT:SIGNALNAME (e.g. ACCELERATOR_RING:TEMP_PROBE_4, although they can be much less verbose to save time).

Most operations are driven directly from a standalone GUI package such as EDM (extensible display manager) or MEDM (Motif/EDM). These allow creation of GUI screens with dials, gauges, text boxes, simple animations, etc. Newer control systems and GUI interfaces such as CSS/BOY are being investigated.

However it is not just GUI software which can interact with EPICS: any software which can speak the CA protocol can get and put values of records. For example, on the EPICS website there are several extension packages which allow CA support in things like MATLAB, LabVIEW, Perl, Python, Tcl, ActiveX, etc. Hence it is easy to do things like write scripts that can activate EPICS controlled equipment.

Record Types

There are different types of records available in EPICS. Here are some common types. Note that in addition to the other records not mentioned here, it is possible to create your own record type to interact with a device.

Each record has various fields in it, which are used for various tasks. AI and AO Analog Input and Output records can obviously store an analog value, and are typically used for things like set-points, temperatures, pressure, flow rates, etc. A limited amount of conversion to and from raw device data is available natively in the record (typically scaling and offsetting, but not advanced conversion like two's complement or logarithmic). BI and BO Binary Input and Output records are generally used for commands and statuses to and from equipment. Calc and Calcout These two records can access other records and perform a calculation based on their values. (E.g. calculate the efficiency of a motor by a function of the current and voltage input and output, and converting to a percentage for the operator to read). Stepper Motor Control of a stepper motor. Allows settings of things like accelerations and velocities, as well as position.

Record Processing

The processing of a record is configured in a number of fields that allow scheduling to be periodic, triggered by a hardware event from the I/O, triggered from another record in a processing chain, or triggered by a read reference from another record or a Channel Access write. With this variety of scan mechanism, control and data acquisition strategies can be configured to minimize latency or assure coherence in a configured processing chain such as a PID (Proportional, Integral, Derivative) feedback loop or interlock.

Facilities Using EPICS

A partial list of facilities using EPICS:

- Australia

    o Australian Synchrotron

    o ANTARES – Australian Nuclear Science and Technology Organisation

    o ASKAP (Australian Square Kilometre Array Pathfinder) – CSIRO

    o Heavy Ion Accelerator at the Australian National University

- Asia
  - KSTAR – Korea Superconducting Tokamak Advanced Research (Republic of Korea)
  - J-PARC – Joint Facility for High Intensity Proton Accelerators (Japan)
  - RIBF – RIKEN RI Beam Factory Project (Japan)
  - BSRF – Beijing Synchrotron Radiation Laboratory (China)
  - VECC – Variable Energy Cyclotron Centre (India)
- Middle-East
  - Synchrotron-Light for Experimental Science and Applications in the Middle East (SESAME) – (Jordan)
- Europe
  - Berliner Elektronenspeicherring für Synchrotronstrahlung (BESSY II) – Helmholtz-Zentrum Berlin (Germany)
  - Deutsches Elektronen Synchrotron (DESY) (Germany)
  - Diamond Light Source – Rutherford Appleton Laboratory (England)
  - European Spallation Source ERIC (ESS) (Sweden)
  - International Thermonuclear Experimental Reactor (ITER) (France)
  - Laboratori Nazionali di Legnaro (Italy)
  - S-DALINAC – Technische Universität Darmstadt (Germany)
  - Spiral2 Système de Production d'Ions RadioActifs en Ligne de deuxième génération (France)
  - Swiss Light Source – Paul Scherrer Institut (Switzerland)
  - GSI/FAIR (Germany)
  - IFMIF – International Fusion Materials Irradiation Facility (Japan, European Union, United States, and Russia)
  - International Muon Ionization Cooling Experiment (MICE) – RAL (UK)
- North America
  - Advanced Light Source – Lawrence Berkeley National Laboratory (United States)
  - Advanced Photon Source – Argonne National Laboratory (United States)
  - Apache Point Observatory (United States)

- o   Canadian Light Source – Saskatoon, Saskatchewan (Canada)

- o   Canadian Neutron Beam Centre – Chalk River Laboratories (Canada)

- o   FNAL – Fermi National Accelerator Laboratory (United States)

- o   Gemini Observatory (United States)

- o   W. M. Keck Observatory (United States)

- o   Laser Interferometer Gravitational-Wave Observatory (LIGO) (United States)

- o   Los Alamos Neutron Science Center – Los Alamos National Laboratory (United States)

- o   National Spherical Torus Experiment – Princeton Plasma Physics Laboratory (United States)

- o   National Spherical Torus Experiment Upgrade – Princeton Plasma Physics Laboratory (United States)

- o   National Superconducting Cyclotron Laboratory – Michigan State University (United States)

- o   National Synchrotron Light Source – Brookhaven National Laboratory (United States)

- o   Spallation Neutron Source – Oak Ridge National Laboratory (United States)

- o   Stanford Synchrotron Radiation Laboratory – Stanford University (United States)

- o   Linac Coherent Light Source – SLAC National Accelerator Laboratory (United States)

- o   TJNAF – Thomas Jefferson National Accelerator Facility (United States)

- o   TRIUMF – Located on the campus of the University of British Columbia (Canada)

- South America

  - o   LNLS – Laboratório Nacional de Luz Síncrotron (Brazil)

- Africa

  - o   iThemba LABS – South Africa – iThemba Laboratory for Accelerator-Based Sciences – Faure, Cape Town (South Africa)

HVAC Control System

HVAC (stands for Heating, Ventilation and Air Conditioning) equipment needs a control system to regulate the operation of a heating and/or air conditioning system. Usually a sensing device is used to compare the actual state (e.g. temperature) with a target state. Then the control system draws a conclusion what action has to be taken (e.g. start the blower).

Direct Digital Control

Central controllers and most terminal unit controllers are programmable, meaning the direct digital control program code may be customized for the intended use. The program features include time schedules, setpoints, controllers, logic, timers, trend logs, and alarms. The unit controllers typically have analog and digital inputs that allow measurement of the variable (temperature, humidity, or pressure) and analog and digital outputs for control of the transport medium (hot/cold water and/or steam). Digital inputs are typically (dry) contacts from a control device, and analog inputs are typically a voltage or current measurement from a variable (temperature, humidity, velocity, or pressure) sensing device. Digital outputs are typically relay contacts used to start and stop equipment, and analog outputs are typically voltage or current signals to control the movement of the medium (air/water/steam) control devices such as valves, dampers, and motors.

Groups of DDC controllers, networked or not, form a layer of system themselves. This "subsystem" is vital to the performance and basic operation of the overall HVAC system. The DDC system is the "brain" of the HVAC system. It dictates the position of every damper and valve in a system. It determines which fans, pumps, and chiller run and at what speed or capacity. With this configurable intelligency in this "brain", we are moving to the concept of building automation.

Building Automation System

More complex HVAC systems can interface to Building Automation System (BAS) to allow the building owners to have more control over the heating or cooling units. The building owner can monitor the system and respond to alarms generated by the system from local or remote locations. The system can be scheduled for occupancy or the configuration can be changed from the BAS. Sometimes the BAS is directly controlling the HVAC components. Depending on the BAS different interfaces can be used.

Today, there are also dedicated gateways that connect advanced VRV / VRF and Split HVAC Systems with Home Automation and BMS (Building Management Systems) controllers for centralized control and monitoring, obviating the need to purchase more complex and expensive HVAC systems. In addition, such gateway solutions are capable of providing remote control operation of all HVAC indoor units over the internet incorporating a simple and friendly user interface.

History

It was natural that the first HVAC controllers would be pneumatic since engineers understood fluid control. Thus, mechanical engineers could use their experience with the properties of steam and air to control the flow of heated or cooled air.

After the control of air flow and temperature was standardized, the use of electromechanical relays in ladder logic to switch dampers became standardized. Eventually, the relays became electronic switches, as transistors eventually could handle greater current loads. By 1985, pneumatic controls could no longer compete with this new technology although pneumatic control systems (sometimes decades old) are still common in many older buildings.

By the year 2000, computerized controllers were common. Today, some of these controllers can even be accessed by web browsers, which need no longer be in the same building as the HVAC equipment. This allows some economies of scale, as a single operations center can easily monitor multiple buildings.

Model Predictive Control

Model predictive control (MPC) is an advanced method of process control that has been in use in the process industries in chemical plants and oil refineries since the 1980s. In recent years it has also been used in power system balancing models. Model predictive controllers rely on dynamic models of the process, most often linear empirical models obtained by system identification. The main advantage of MPC is the fact that it allows the current timeslot to be optimized, while keeping future timeslots in account. This is achieved by optimizing a finite time-horizon, but only implementing the current timeslot. MPC has the ability to anticipate future events and can take control actions accordingly. PID and LQR controllers do not have this predictive ability. MPC is nearly universally implemented as a digital control, although there is research into achieving faster response times with specially designed analog circuitry.

Overview

The models used in MPC are generally intended to represent the behavior of complex dynamical systems. The additional complexity of the MPC control algorithm is not generally needed to provide adequate control of simple systems, which are often controlled well by generic PID controllers. Common dynamic characteristics that are difficult for PID controllers include large time delays and high-order dynamics.

MPC models predict the change in the dependent variables of the modeled system that will be caused by changes in the independent variables. In a chemical process, independent variables that can be adjusted by the controller are often either the setpoints of regulatory PID controllers (pressure, flow, temperature, etc.) or the final control element (valves, dampers, etc.). Independent variables that cannot be adjusted by the controller are used as disturbances. Dependent variables in these processes are other measurements that represent either control objectives or process constraints.

MPC uses the current plant measurements, the current dynamic state of the process, the MPC models, and the process variable targets and limits to calculate future changes in the dependent variables. These changes are calculated to hold the dependent variables close to target while honoring constraints on both independent and dependent variables. The MPC typically sends out only the first change in each independent variable to be implemented, and repeats the calculation when the next change is required.

While many real processes are not linear, they can often be considered to be approximately linear over a small operating range. Linear MPC approaches are used in the majority of applications with the feedback mechanism of the MPC compensating for prediction errors due to structural mismatch between the model and the process. In model predictive controllers that consist only of linear models, the superposition principle of linear algebra enables the effect of changes in multiple independent variables to be added together to predict the response of the dependent variables. This simplifies the control problem to a series of direct matrix algebra calculations that are fast and robust.

When linear models are not sufficiently accurate to represent the real process nonlinearities, several approaches can be used. In some cases, the process variables can be transformed before and/or after the linear MPC model to reduce the nonlinearity. The process can be controlled with nonlinear MPC that uses a nonlinear model directly in the control application. The nonlinear model may be in the form of an empirical data fit (e.g. artificial neural networks) or a high-fidelity dynamic model based on fundamental mass and energy balances. The nonlinear model may be linearized to derive a Kalman filter or specify a model for linear MPC.

An algorithmic study by El-Gherwi, Budman, and El Kamel shows that utilizing a dual-mode approach can provide significant reduction in online computations while maintaining comparative performance to a non-altered implementation. The proposed algorithm solves N convex optimization problems in parallel based on exchange of information among controllers.

Theory Behind MPC

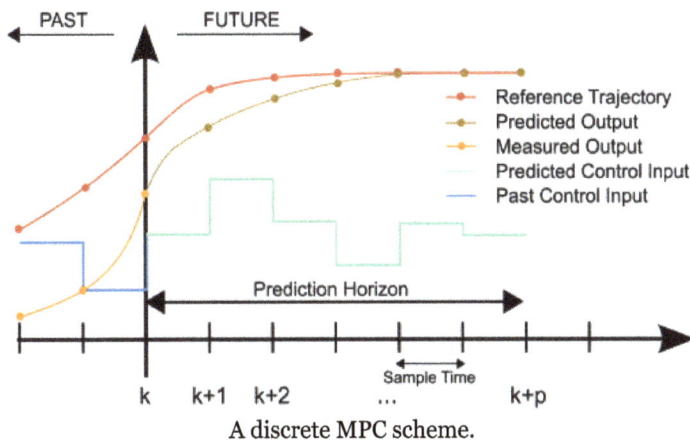

A discrete MPC scheme.

MPC is based on iterative, finite-horizon optimization of a plant model. At time t the current plant state is sampled and a cost minimizing control strategy is computed (via a numerical minimization algorithm) for a relatively short time horizon in the future: $[t, t+T]$. Specifically, an online or on-the-fly calculation is used to explore state trajectories that emanate from the current state and find (via the solution of Euler–Lagrange equations) a cost-minimizing control strategy until time $t+T$. Only the first step of the control strategy is implemented, then the plant state is sampled again and the calculations are repeated starting from the new current state, yielding a new control and new predicted state path. The prediction horizon keeps being shifted forward and for this reason MPC is also called receding horizon control. Although this approach is not optimal, in practice it has given very good results. Much academic research has been done to find fast methods of solution of Euler–Lagrange type equations, to understand the global stability properties of MPC's local optimization, and in general to improve the MPC method. To some extent the theoreticians have been trying to catch up with the control engineers when it comes to MPC.

Principles of MPC

Model Predictive Control (MPC) is a multivariable control algorithm that uses:

- an internal dynamic model of the process

- a history of past control moves and

- an optimization cost function J over the receding prediction horizon,

to calculate the optimum control moves.

An example of a non-linear cost function for optimization is given by:

$$J = \sum_{i=1}^{N} w_{x_i} (r_i - x_i)^2 + \sum_{i=1}^{N} w_{u_i} \Delta u_i^2$$

without violating constraints (low/high limits)

With:

$x_i = i$ -th controlled variable (e.g. measured temperature)

$r_i = i$ -th reference variable (e.g. required temperature)

$u_i = i$ -th manipulated variable (e.g. control valve)

w_{x_i} = weighting coefficient reflecting the relative importance of x_i

w_{u_i} = weighting coefficient penalizing relative big changes in u_i

etc.

Nonlinear MPC

Nonlinear Model Predictive Control, or NMPC, is a variant of model predictive control (MPC) that is characterized by the use of nonlinear system models in the prediction. As in linear MPC, NMPC requires the iterative solution of optimal control problems on a finite prediction horizon. While these problems are convex in linear MPC, in nonlinear MPC they are not convex anymore. This poses challenges for both NMPC stability theory and numerical solution.

The numerical solution of the NMPC optimal control problems is typically based on direct optimal control methods using Newton-type optimization schemes, in one of the variants: direct single shooting, direct multiple shooting methods, or direct collocation. NMPC algorithms typically exploit the fact that consecutive optimal control problems are similar to each other.

This allows to initialize the Newton-type solution procedure efficiently by a suitably shifted guess from the previously computed optimal solution, saving considerable amounts of computation time. The similarity of subsequent problems is even further exploited by path following algorithms (or "real-time iterations") that never attempt to iterate any optimization problem to convergence, but instead only take one iteration towards the solution of the most current NMPC problem, before proceeding to the next one, which is suitably initialized.

While NMPC applications have in the past been mostly used in the process and chemical industries with comparatively slow sampling rates, NMPC is more and more being applied to applications with high sampling rates, e.g., in the automotive industry, or even when the states are distributed in space (Distributed parameter systems). As an application in aerospace, recently, NMPC has been used to track optimal terrain-following/avoidance trajectories in real-time.

Robust MPC

Robust variants of Model Predictive Control (MPC) are able to account for set bounded disturbance while still ensuring state constraints are met. There are three main approaches to robust MPC:

- *Min-max MPC.* In this formulation, the optimization is performed with respect to all possible evolutions of the disturbance. This is the optimal solution to linear robust control problems, however it carries a high computational cost.

- *Constraint Tightening MPC.* Here the state constraints are enlarged by a given margin so that a trajectory can be guaranteed to be found under any evolution of disturbance.

- *Tube MPC.* This uses an independent nominal model of the system, and uses a feedback controller to ensure the actual state converges to the nominal state. The amount of separation required from the state constraints is determined by the robust positively invariant (RPI) set, which is the set of all possible state deviations that may be introduced by disturbance with the feedback controller.

- *Multi-stage MPC.* This uses a scenario-tree formulation by approximating the uncertainty space with a set of samples and the approach is non-conservative because it takes into account that the measurement information is available at every time stages in the prediction and the decisions at every stage can be different and can act as recourse to counteract the effects of uncertainties. The drawback of the approach however is that the size of the problem grows exponentially with the number of uncertainties and the prediction horizon.

Commercially Available MPC Software

Commercial MPC packages are available and typically contain tools for model identification and analysis, controller design and tuning, as well as controller performance evaluation.

A survey of commercially available packages has been provided by S.J. Qin and T.A. Badgwell in *Control Engineering Practice* 11 (2003) 733–764.

Nyquist Stability Criterion

In control theory and stability theory, the Nyquist stability criterion, discovered by Swedish-American electrical engineer Harry Nyquist at Bell Telephone Laboratories in 1932, is a graphical technique for determining the stability of a dynamical system. Because it only looks at the Nyquist plot of the open loop systems, it can be applied without explicitly computing the poles and zeros of either the closed-loop or open-loop system (although the number of each type of right-half-plane singularities must be known). As a result, it can be applied to systems defined by non-rational functions, such as systems with delays. In contrast to Bode plots, it can handle transfer functions with right half-plane singularities. In addition, there is a natural generalization to more complex systems with multiple inputs and multiple outputs, such as control systems for airplanes.

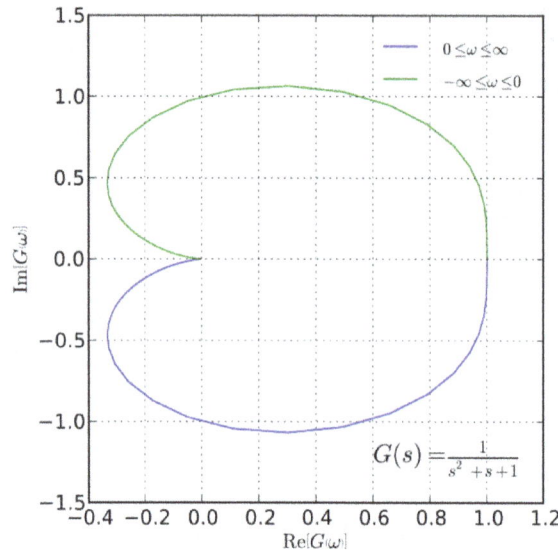

The Nyquist plot for $G(s) = \dfrac{1}{s^2 + s + 1}$.

The Nyquist criterion is widely used in electronics and control system engineering, as well as other fields, for designing and analyzing systems with feedback. While Nyquist is one of the most general stability tests, it is still restricted to linear, time-invariant (LTI) systems. Non-linear systems must use more complex stability criteria, such as Lyapunov or the circle criterion. While Nyquist is a graphical technique, it only provides a limited amount of intuition for why a system is stable or unstable, or how to modify an unstable system to be stable. Techniques like Bode plots, while less general, are sometimes a more useful design tool.

Background

We consider a system whose open loop transfer function (OLTF) is $G(s)$; when placed in a closed loop with negative feedback $H(s)$, the closed loop transfer function (CLTF) then becomes $\dfrac{G}{1+GH}$. Stability can be determined by examining the roots of the desensitivity factor polynomial $1+GH$, e.g. using the Routh array, but this method is somewhat tedious. Conclusions can also be reached by examining the OLTF, using its Bode plots or, as here, polar plot of the OLTF using the Nyquist criterion, as follows.

Any Laplace domain transfer function $T(s)$ can be expressed as the ratio of two polynomials:

$$T(s) = \frac{N(s)}{D(s)}.$$

The roots of $N(s)$ are called the *zeros* of $T(s)$, and the roots of $D(s)$ are the *poles* of $T(s)$. The poles of $T(s)$ are also said to be the roots of the "characteristic equation" $D(s) = 0$.

The stability of $T(s)$ is determined by the values of its poles: for stability, the real part of every pole must be negative. If $T(s)$ is formed by closing a negative unity feedback loop around the open-loop transfer function

$G(s) \quad \dfrac{A(s)}{B(s)}$, then the roots of the characteristic equation are also the zeros of $1 + G(s)$,

or simply the roots of $A(s) + B(s) = 0$.

Cauchy's Argument Principle

From complex analysis, specifically the argument principle, we know that a contour Γ_s drawn in the complex $F(s)$ plane, encompassing but not passing through any number of zeros and poles of a function $F(s)$, can be mapped to another plane (the $F(s)$ plane) by the function $F(s)$. The Nyquist plot of $F(s)$, which is the contour $\Gamma_{F(s)} = F(\Gamma_s)$ will encircle the point $s = -1/k$ of the $F(s)$ plane N times, where $N = Z - P$. Here are Z and P respectively the number of zeros of $1 + kF(s)$ and poles of $F(s)$ inside the contour Γ_s. Note that we count encirclements in the $F(s)$ plane in the same sense as the contour Γ_s and that encirclements in the opposite direction are *negative* encirclements. That is, we consider clockwise encirclements to be negative and counterclockwise encirclements to be positive.

Instead of Cauchy's argument principle, the original paper by Harry Nyquist in 1932 uses a less elegant approach. The approach explained here is similar to the approach used by Leroy MacColl (Fundamental theory of servomechanisms 1945) or by Hendrik Bode (Network analysis and feedback amplifier design 1945), both of whom also worked for Bell Laboratories. This approach appears in most modern textbooks on control theory.

The Nyquist Criterion

We first construct the Nyquist contour, a contour that encompasses the right-half of the complex plane:

- a path traveling up the $j\omega$ axis, from $0 - j\infty$ to $0 + j\infty$.

- a semicircular arc, with radius $r \to \infty$,, that starts at $0 + j\infty$ and travels clock-wise to $0 - j\infty$.

The Nyquist contour mapped through the function $1 + G(s)$ yields a plot of $1 + G(s)$ in the complex plane. By the Argument Principle, the number of clock-wise encirclements of the origin must be the number of zeros of $1 + G(s)$ in the right-half complex plane minus the poles of $1 + G(s)$ in the right-half complex plane. If instead, the contour is mapped through the open-loop transfer function $G(s)$, the result is the Nyquist Plot of $G(s)$. By counting the resulting contour's encirclements of -1, we find the difference between the number of poles and zeros in the right-half complex plane of $1 + G(s)$. Recalling that the zeros of $1 + G(s)$ are the poles of the closed-loop system, and noting that the poles of $1 + G(s)$ are same as the poles of $G(s)$, we now state The Nyquist Criterion:

Given a Nyquist contour Γ_s, let P be the number of poles of $G(s)$ encircled by $\tilde{A}_s$, and be Z the number of zeros of $1+G(s)$ encircled by Γ_s. Alternatively, and more importantly, Z is the number of poles of the closed loop system in the right half plane. The resultant contour in the $G(s)$-plane, $\Gamma_{G(s)}$ shall encircle (clock-wise) the point $(-1+j0)N$ times such that $N = Z - P$. If the system is originally open-loop unstable, feedback is necessary to stabilize the system. Right-half-plane (RHP) poles represent that instability. For closed-loop stability of a system, the number of closed-loop roots in the right half of the s-plane must be zero. Hence, the number of counter-clockwise encirclements about $-1+j0$ must be equal to the number of open-loop poles in the RHP. Any clockwise encirclements of the critical point by the open-loop frequency response (when judged from low frequency to high frequency) would indicate that the feedback control system would be destabilizing if the loop were closed. (Using RHP zeros to "cancel out" RHP poles does not remove the instability, but rather ensures that the system will remain unstable even in the presence of feedback, since the closed-loop roots travel between open-loop poles and zeros in the presence of feedback. In fact, the RHP zero can make the unstable pole unobservable and therefore not stabilizable through feedback.)

The Nyquist Criterion for Systems with Poles on the Imaginary Axis

The above consideration was conducted with an assumption that the open-loop transfer function $G(s)$ does not have any pole on the imaginary axis (i.e. poles of the form $0+j\omega$). This results from the requirement of the argument principle that the contour cannot pass through any pole of the mapping function. The most common case are systems with integrators (poles at zero).

To be able to analyze systems with poles on the imaginary axis, the Nyquist Contour can be modified to avoid passing through the point $0+j\omega$. One way to do it is to construct a semicircular arc with radius $r \to 0$ around $0+j\omega$, that starts at $0+j(\omega-r)$ and travels anticlockwise to $0+j(\omega+r)$. Such a modification implies that the phasor $G(s)$ travels along an arc of infinite radius by $-l\pi$, where l is the multiplicity of the pole on the imaginary axis.

Mathematical Derivation

Our goal is to, through this process, check for the stability of the transfer function of our unity feedback system with gain k, which is given by

$$T(s) = \frac{kG(s)}{1+kG(s)}$$

That is, we would like to check whether the characteristic equation of the above transfer function, given by

$$D(s) = 1 + kG(s) = 0$$

has zeros outside the open left-half-plane (commonly initialized as the OLHP).

We suppose that we have a clockwise (i.e. negatively oriented) contour Γ_s enclosing the right hand plane, with indentations as needed to avoid passing through zeros or poles of the function $G(s)$. Cauchy's argument principle states that

$$-\frac{1}{2\pi i}\oint_{\Gamma_s}\frac{D'(s)}{D(s)}ds = N = Z - P$$

Where Z denotes the number of zeros of $D(s)$ enclosed by the contour and P denotes the number of poles of $D(s)$ by the same contour. Rearranging, we have $Z = N + P$, which is to say

$$Z = -\frac{1}{2\pi i}\oint_{\Gamma_s}\frac{D'(s)}{D(s)}ds + P$$

We then note that $D(s) = 1 + kG(s)$ has exactly the same poles as $G(s)$. Thus, we may find P by counting the poles of $G(s)$ that appear within the contour, that is, within the open right half plane (ORHP).

We will now rearrange the above integral via substitution. That is, setting $u(s) = D(s)$, we have

$$N = -\frac{1}{2\pi i}\oint_{\Gamma_s}\frac{D'(s)}{D(s)}ds = -\frac{1}{2\pi i}\oint_{u(\Gamma_s)}\frac{1}{u}du$$

We then make a further substitution, setting $v(u) = \dfrac{u-1}{k}$. This gives us

$$N = -\frac{1}{2\pi i}\oint_{u(\Gamma_s)}\frac{1}{u}du = -\frac{1}{2\pi i}\oint_{v(u(\Gamma_s))}\frac{1}{v+1/k}dv$$

We now note that $v(u(\Gamma_s)) = \dfrac{D(\Gamma_s)-1}{k} = G(\Gamma_s)$ gives us the image of our contour under $G(s)$,

which is to say our Nyquist Plot. We may further reduce the integral

$$N = -\frac{1}{2\pi i}\oint_{G(\Gamma_s)}\frac{1}{v+1/k}dv$$

by applying Cauchy's integral formula. In fact, we find that the above integral corresponds precisely to the number of times the Nyquist Plot encircles the point $-1/k$ clockwise. Thus, we may finally state that

$Z = N + P = $ (number of times the Nyquist plot encircles -1/k clockwise) + (number of poles of G(s) in ORHP)

We thus find that $T(s)$ as defined above corresponds to a stable unity-feedback system when Z, as evaluated above, is equal to 0.

Summary

- If the open-loop transfer function $G(s)$ has a zero pole of multiplicity l, then the Nyquist plot has a discontinuity at $\omega = 0$. During further analysis it should be assumed that the phasor travels l times clock-wise along a semicircle of infinite radius. After applying this rule, the zero poles should be neglected, i.e. if there are no other unstable poles, then the open-loop transfer function $G(s)$ should be considered stable.

- If the open-loop transfer function $G(s)$ is stable, then the closed-loop system is unstable for *any* encirclement of the point -1.

- If the open-loop transfer function $G(s)$ is *unstable*, then there must be one *counter* clock-wise encirclement of -1 for each pole of $G(s)$ in the right-half of the complex plane.

- The number of surplus encirclements (greater than N+P) is exactly the number of unstable poles of the closed-loop system.

- However, if the graph happens to pass through the point $-1+j0$, then deciding upon even the marginal stability of the system becomes difficult and the only conclusion that can be drawn from the graph is that there exist zeros on the $j\omega$ axis.

Electric Power System

A steam turbine used to provide electric power.

An electric power system is a network of electrical components used to supply, transfer and use electric power. An example of an electric power system is the network that supplies a region's homes and industry with power—for sizeable regions, this power system is known as *the grid* and can be broadly divided into the generators that supply the power, the transmission system that carries the power from the generating centres to the load centres and the distribution system that feeds the power to nearby homes and industries. Smaller power systems are also found in industry, hospitals, commercial buildings and homes. The majority of these systems rely upon three-phase AC power—the standard for large-scale power transmission and distribution across the modern world. Specialised power systems that do not always rely upon three-phase AC power are found in aircraft, electric rail systems, ocean liners and automobiles.

History

In 1881 two electricians built the world's first power system at Godalming in England. It was powered by a power station consisting of two waterwheels that produced an alternating current that in turn supplied seven Siemens arc lamps at 250 volts and 34 incandescent lamps at 40 volts. However supply to the lamps was intermittent and in 1882 Thomas Edison and his company, The Edison Electric Light Company, developed the first steam powered electric power station on Pearl Street in New York City. The Pearl Street Station initially powered around 3,000 lamps for 59 customers. The power station used direct current and operated at a single voltage. Direct current power could not be easily transformed to the higher voltages necessary to minimise power loss during long-distance transmission, so the maximum economic distance between the generators and load was limited to around half-a-mile (800 m).

A sketch of the Pearl Street Station

That same year in London Lucien Gaulard and John Dixon Gibbs demonstrated the first transformer suitable for use in a real power system. The practical value of Gaulard and Gibbs' transformer was demonstrated in 1884 at Turin where the transformer was used to light up forty kilometres (25 miles) of railway from a single alternating current generator. Despite the success of the system, the pair made some fundamental mistakes. Perhaps the most serious was connecting the primaries of the transformers in series so that active lamps would affect the brightness of other lamps further down the line.

In 1885 George Westinghouse, an American entrepreneur, obtained the patent rights to the Gaulard Gibbs transformer and imported a number of them along with a Siemens generator and set his engineers to experimenting with them in the hopes of improving them for use in a commercial power system. One of Westinghouse's engineers, William Stanley, recognised the problem with connecting transformers in series as opposed to parallel and also realised that making the iron core of a transformer a fully enclosed loop would improve the voltage regulation of the secondary winding. Using this knowledge he built the first practical transformer based alternating current power system at Great Barrington, Massachusetts in 1886. Westinghouse would begin installing multi-voltage AC transformer systems in competition with the Edison company later that year. In 1888 Westinghouse would also licensed Nikola Tesla's US patents for a polyphase AC induction motor and transformer designs and hired Tesla for one year to be a consultant at the Westinghouse Electric & Manufacturing Company's Pittsburgh labs.

By 1888 the electric power industry was flourishing, and power companies had built thousands of power systems (both direct and alternating current) in the United States and Europe. These networks were effectively dedicated to providing electric lighting. During this time the rivalry between Thomas Edison and George Westinghouse's companies had grown into propaganda campaign over which form of transmission (direct or alternating current) was superior, a searies of events known as the "War of Currents". In 1891, Westinghouse installed the first major power system that was designed to drive a 100 horsepower (75 kW) synchronous electric motor, not just provide electric lighting, at Telluride, Colorado. On the other side of the Atlantic, Mikhail Dolivo-Dobrovolsky built a 20 kV 176 km three-phase transmission line from Lauffen am Neckar to Frankfurt am Main for the Electrical Engineering Exhibition in Frankfurt. In the US the AC/DC competition came to the end when Edison General Electric was taken over by their chief AC rival, the Thomson-Houston Electric Company, forming General Electric. In 1895, after a protracted decision-making process, alternating current was chosen as the transmission standard with Westinghouse building the Adams No. 1 generating station at Niagara Falls and General Electric building the three-phase alternating current power system to supply Buffalo at 11 kV.

Developments in power systems continued beyond the nineteenth century. In 1936 the first experimental HVDC (high voltage direct current) line using mercury arc valves was built between Schenectady and Mechanicville, New York. HVDC had previously been achieved by series-connected direct current generators and motors (the Thury system) although this suffered from serious reliability issues. In 1957 Siemens demonstrated the first solid-state rectifier, but it was not until the early 1970s that solid-state devices became the standard in HVDC. In recent times, many important developments have come from extending innovations in the ICT field to the power engineering field. For example, the development of computers meant load flow studies could be run more efficiently allowing for much better planning of power systems. Advances in information technology and telecommunication also allowed for remote control of a power system's switchgear and generators.

Basics of Electric Power

An external AC to DC power adapter used for household appliances

Electric power is the product of two quantities: current and voltage. These two quantities can vary with respect to time (AC power) or can be kept at constant levels (DC power).

Most refrigerators, air conditioners, pumps and industrial machinery use AC power whereas most computers and digital equipment use DC power (the digital devices you plug into the mains typically have an internal or external power adapter to convert from AC to DC power). AC power has the advantage of being easy to transform between voltages and is able to be generated and utilised by brushless machinery. DC power remains the only practical choice in digital systems and can be more economical to transmit over long distances at very high voltages.

The ability to easily transform the voltage of AC power is important for two reasons: Firstly, power can be transmitted over long distances with less loss at higher voltages. So in power systems where generation is distant from the load, it is desirable to step-up (increase) the voltage of power at the generation point and then step-down (decrease) the voltage near the load. Secondly, it is often more economical to install turbines that produce higher voltages than would be used by most appliances, so the ability to easily transform voltages means this mismatch between voltages can be easily managed.

Solid state devices, which are products of the semiconductor revolution, make it possible to transform DC power to different voltages, build brushless DC machines and convert between AC and DC power. Nevertheless, devices utilising solid state technology are often more expensive than their traditional counterparts, so AC power remains in widespread use.

Balancing the Grid

One of the main difficulties in power systems is that the amount of active power consumed plus losses should always equal the active power produced. If more power would be produced than consumed the frequency would rise and vice versa. Even small deviations from the nominal frequency value would damage synchronous machines and other appliances. Making sure the frequency is constant is usually the task of a transmission system operator. In some countries (for example in the European Union) this is achieved through a balancing market using ancillary services.

Components of Power Systems

Supplies

The majority of the world's power still comes from coal-fired power stations like this.

All power systems have one or more sources of power. For some power systems, the source of power is external to the system but for others it is part of the system itself—it is these internal power sources that are discussed in the remainder of this section. Direct current power can be supplied by batteries, fuel cells or photovoltaic cells. Alternating current power is typically supplied by a rotor that spins in a magnetic field in a device known as a turbo generator. There have been a wide range of techniques used to spin a turbine's rotor, from steam heated using fossil fuel (including coal, gas and oil) or nuclear energy, falling water (hydroelectric power) and wind (wind power).

The speed at which the rotor spins in combination with the number of generator poles determines the frequency of the alternating current produced by the generator. All generators on a single synchronous system, for example the national grid, rotate at sub-multiples of the same speed and so generate electric current at the same frequency. If the load on the system increases, the generators will require more torque to spin at that speed and, in a typical power station, more steam must be supplied to the turbines driving them. Thus the steam used and the fuel expended are directly dependent on the quantity of electrical energy supplied. An exception exists for generators incorporating power electronics such as gearless wind turbines or linked to a grid through an asynchronous tie such as a HVDC link — these can operate at frequencies independent of the power system frequency.

Depending on how the poles are fed, alternating current generators can produce a variable number of phases of power. A higher number of phases leads to more efficient power system operation but also increases the infrastructure requirements of the system.

Electricity grid systems connect multiple generators and loads operating at the same frequency and number of phases, the commonest being three-phase at 50 or 60 Hz. However, there are other considerations. These range from the obvious: How much power should the generator be able to supply? What is an acceptable length of time for starting the generator (some generators can take hours to start)? Is the availability of the power source acceptable (some renewables are only available when the sun is shining or the wind is blowing)? To the more technical: How should the generator start (some turbines act like a motor to bring themselves up to speed in which case they need an appropriate starting circuit)? What is the mechanical speed of operation for the turbine and consequently what are the number of poles required? What type of generator is suitable (synchronous or asynchronous) and what type of rotor (squirrel-cage rotor, wound rotor, salient pole rotor or cylindrical rotor)?

Loads

A toaster is great example of a single-phase load that might appear in a residence. Toasters typically draw 2 to 10 amps at 110 to 260 volts consuming around 600 to 1200 watts of power

Power systems deliver energy to loads that perform a function. These loads range from household appliances to industrial machinery. Most loads expect a certain voltage and, for alternating current devices, a certain frequency and number of phases. The appliances found in your home, for example, will typically be single-phase operating at 50 or 60 Hz with a voltage between 110 and 260 volts (depending on national standards). An exception exists for centralized air conditioning systems as these are now typically three-phase because this allows them to operate more efficiently. All devices in your house will also have a wattage, this specifies the amount of power the device consumes. At any one time, the net amount of power consumed by the loads on a power system must equal the net amount of power produced by the supplies less the power lost in transmission.

Making sure that the voltage, frequency and amount of power supplied to the loads is in line with expectations is one of the great challenges of power system engineering. However it is not the only challenge, in addition to the power used by a load to do useful work (termed real power) many alternating current devices also use an additional amount of power because they cause the alternating voltage and alternating current to become slightly out-of-sync (termed reactive power). The reactive power like the real power must balance (that is the reactive power produced on a system must equal the reactive power consumed) and can be supplied from the generators, however it is often more economical to supply such power from capacitors.

A final consideration with loads is to do with power quality. In addition to sustained overvoltages and undervoltages (voltage regulation issues) as well as sustained deviations from the system frequency (frequency regulation issues), power system loads can be adversely affected by a range of temporal issues. These include voltage sags, dips and swells, transient overvoltages, flicker, high frequency noise, phase imbalance and poor power factor. Power quality issues occur when the power supply to a load deviates from the ideal: For an AC supply, the ideal is the current and voltage in-sync fluctuating as a perfect sine wave at a prescribed frequency with the voltage at a prescribed amplitude. For DC supply, the ideal is the voltage not varying from a prescribed level. Power quality issues can be especially important when it comes to specialist industrial machinery or hospital equipment.

Conductors

Conductors carry power from the generators to the load. In a grid, conductors may be classified as belonging to the transmission system, which carries large amounts of power at high voltages (typically more than 69 kV) from the generating centres to the load centres, or the distribution system, which feeds smaller amounts of power at lower voltages (typically less than 69 kV) from the load centres to nearby homes and industry.

Choice of conductors is based upon considerations such as cost, transmission losses and other desirable characteristics of the metal like tensile strength. Copper, with lower resistivity than Aluminum, was the conductor of choice for most power systems. However, Aluminum has lower cost for the same current carrying capacity and is the primary metal used for transmission line conductors. Overhead line conductors may be reinforced with steel or aluminum alloys.

Conductors in exterior power systems may be placed overhead or underground. Overhead conductors are usually air insulated and supported on porcelain, glass or polymer insulators. Cables used for underground transmission or building wiring are insulated with cross-linked polyethylene or other flexible insulation. Large conductors are stranded for ease of handling; small conductors used for building wiring are often solid, especially in light commercial or residential construction.

Conductors are typically rated for the maximum current that they can carry at a given temperature rise over ambient conditions. As current flow increases through a conductor it heats up. For insulated conductors, the rating is determined by the insulation. For overhead conductors, the rating is determined by the point at which the sag of the conductors would become unacceptable.

Capacitors and Reactors

The majority of the load in a typical AC power system is inductive; the current lags behind the voltage. Since the voltage and current are out-of-phase, this leads to the emergence of an "imaginary" form of power known as reactive power. Reactive power does no measurable work but is transmitted back and forth between the reactive power source and load every cycle. This reactive power can be provided by the generators themselves, through the adjustment of generator excitation, but it is often cheaper to provide it through capacitors, hence capacitors are often placed near inductive loads to reduce current demand on the power system (i.e., increase the power factor), which may never exceed 1.0, and which represents a purely resistive load. Power factor correction may be applied at a central substation, through the use of so-called "synchronous condensers" (synchronous machines which act as condensers which are variable in VAR value, through the adjustment of machine excitation) or adjacent to large loads, through the use of so-called "static condensers" (condensers which are fixed in VAR value).

Reactors consume reactive power and are used to regulate voltage on long transmission lines. In light load conditions, where the loading on transmission lines is well below the surge impedance loading, the efficiency of the power system may actually be improved by switching in reactors. Reactors installed in series in a power system also limit rushes of current flow, small reactors are therefore almost always installed in series with capacitors to limit the current rush associated with switching in a capacitor. Series reactors can also be used to limit fault currents.

Capacitors and reactors are switched by circuit breakers, which results in moderately large steps in reactive power. A solution comes in the form of static VAR compensators and static synchronous compensators. Briefly, static VAR compensators work by switching in capacitors using thyristors as opposed to circuit breakers allowing capacitors to be switched-in and switched-out within a single cycle. This provides a far more refined response than circuit breaker switched capacitors. Static synchronous compensators take a step further by achieving reactive power adjustments using only power electronics.

Power Electronics

Power electronics are semi-conductor based devices that are able to switch quantities of power ranging from a few hundred watts to several hundred megawatts. Despite their relatively simple function, their speed of operation (typically in the order of nanoseconds) means they are capable of a wide range of tasks that would be difficult or impossible with conventional technology. The classic function of power electronics is rectification, or the conversion of AC-to-DC power, power electronics are therefore found in almost every digital device that is supplied from an AC source either as an adapter that plugs into the wall or as component internal to the device. High-powered power electronics can also be used to convert AC power to DC power for long distance transmission in a system known as HVDC.

HVDC is used because it proves to be more economical than similar high voltage AC systems for very long distances (hundreds to thousands of kilometres). HVDC is also desirable for interconnects because it allows frequency independence thus improving system stability. Power electronics are also essential for any power source that is required to produce an AC output but that by its nature produces a DC output. They are therefore used by many photovoltaic installations both industrial and residential.

Power electronics also feature in a wide range of more exotic uses. They are at the heart of all modern electric and hybrid vehicles—where they are used for both motor control and as part of the brushless DC motor. Power electronics are also found in practically all modern petrol-powered vehicles, this is because the power provided by the car's batteries alone is insufficient to provide ignition, air-conditioning, internal lighting, radio and dashboard displays for the life of the car. So the batteries must be recharged while driving using DC power from the engine—a feat that is typically accomplished using power electronics. Whereas conventional technology would be unsuitable for a modern electric car, commutators can and have been used in petrol-powered cars, the switch to alternators in combination with power electronics has occurred because of the improved durability of brushless machinery.

Some electric railway systems also use DC power and thus make use of power electronics to feed grid power to the locomotives and often for speed control of the locomotive's motor. In the middle twentieth century, rectifier locomotives were popular, these used power electronics to convert AC power from the railway network for use by a DC motor. Today most electric locomotives are supplied with AC power and run using AC motors, but still use power electronics to provide suitable motor control. The use of power electronics to assist with motor control and with starter circuits cannot be underestimated and, in addition to rectification, is responsible for power electronics appearing in a wide range of industrial machinery. Power electronics even appear in modern residential air conditioners.

Power electronics are also at the heart of the variable speed wind turbine. Conventional wind turbines require significant engineering to ensure they operate at some ratio of the system frequency, however by using power electronics this requirement can be eliminated leading to quieter, more flexible and (at the moment) more costly wind turbines. A final example of one of the more exotic uses of power electronics comes from the previous section where the fast-switching times of power electronics were used to provide more refined reactive compensation to the power system.

Protective Devices

Power systems contain protective devices to prevent injury or damage during failures. The quintessential protective device is the fuse. When the current through a fuse exceeds a certain threshold, the fuse element melts, producing an arc across the resulting gap that is then extinguished, interrupting the circuit. Given that fuses can be built as the weak point of a system, fuses are ideal for protecting circuitry from damage. Fuses however have two problems: First, after they have functioned, fuses must be replaced as they cannot be reset. This can prove inconvenient if the fuse is at a remote site or a spare fuse is not on hand. And second, fuses are typically inadequate as the sole safety device in most power systems as they allow current flows well in excess of that that would prove lethal to a human or animal.

The first problem is resolved by the use of circuit breakers—devices that can be reset after they have broken current flow. In modern systems that use less than about 10 kW, miniature circuit breakers are typically used. These devices combine the mechanism that initiates the trip (by sensing excess current) as well as the mechanism that breaks the current flow in a single unit. Some miniature circuit breakers operate solely on the basis of electromagnetism. In these miniature circuit breakers, the current is run through a solenoid, and, in the event of excess current flow, the magnetic pull of the solenoid is sufficient to force open the circuit breaker's contacts (often indirectly through a tripping mechanism). A better design however arises by inserting a bimetallic strip before the solenoid—this means that instead of always producing a magnetic force, the solenoid only produces a magnetic force when the current is strong enough to deform the bimetallic strip and complete the solenoid's circuit.

In higher powered applications, the protective relays that detect a fault and initiate a trip are separate from the circuit breaker. Early relays worked based upon electromagnetic principles similar to those mentioned in the previous paragraph, modern relays are application-specific computers that determine whether to trip based upon readings from the power system. Different relays will initiate trips depending upon different protection schemes. For example, an overcurrent relay might initiate a trip if the current on any phase exceeds a certain threshold whereas a set of differential relays might initiate a trip if the sum of currents between them indicates there may be current leaking to earth. The circuit breakers in higher powered applications are different too. Air is typically no longer sufficient to quench the arc that forms when the contacts are forced open so a variety of techniques are used. One of the most popular techniques is to keep the chamber enclosing the contacts flooded with sulfur hexafluoride (SF_6)—a non-toxic gas that has sound arc-quenching properties. Other techniques are discussed in the reference.

The second problem, the inadequacy of fuses to act as the sole safety device in most power systems, is probably best resolved by the use of residual current devices (RCDs). In any properly functioning electrical appliance the current flowing into the appliance on the active line should equal the current flowing out of the appliance on the neutral line. A residual current device works by monitoring the active and neutral lines and tripping the active line if it notices a difference. Residual current devices require a separate neutral line for each phase and to be able to trip within a time frame before harm occurs. This is typically not a problem in most residential applications where standard wiring provides an active and neutral line for each appliance (that's why your power plugs always have at least two tongs) and the voltages are relatively low however these issues do limit the effectiveness of RCDs in other applications such as industry. Even with the installation of an RCD, exposure to electricity can still prove lethal.

SCADA Systems

In large electric power systems, Supervisory Control And Data Acquisition (SCADA) is used for tasks such as switching on generators, controlling generator output and switching in or out system elements for maintenance. The first supervisory control systems implemented consisted of a panel of lamps and switches at a central console near the controlled plant. The lamps provided feedback on the state of plant (the data acquisition function) and the switches allowed adjustments to the plant to be made (the supervisory control function). Today, SCADA systems are much more sophisticated and, due to advances in communication systems, the consoles controlling the plant no longer need to be near the plant itself. Instead it is now common for plants to be controlled with equipment similar (if not identical) to a desktop computer. The ability to control such plants through computers has increased the need for security—there have already been reports of cyber-attacks on such systems causing significant disruptions to power systems.

Power Systems in Practice

Despite their common components, power systems vary widely both with respect to their design and how they operate. This section introduces some common power system types and briefly explains their operation.

Residential Power Systems

Residential dwellings almost always take supply from the low voltage distribution lines or cables that run past the dwelling. These operate at voltages of between 110 and 260 volts (phase-to-earth) depending upon national standards. A few decades ago small dwellings would be fed a single phase using a dedicated two-core service cable (one core for the active phase and one core for the neutral return). The active line would then be run through a main isolating switch in the fuse box and then split into one or more circuits to feed lighting and appliances inside the house. By convention, the lighting and appliance circuits are kept separate so the failure of an appliance does not leave the dwelling's occupants in the dark. All circuits would be fused with an appropriate fuse based upon the wire size used for that circuit. Circuits would have both an active and neutral wire with both the lighting and power sockets being connected in parallel. Sockets would also be provided with a protective earth. This would be made available to appliances to connect to any metallic casing. If this casing were to become live, the theory is the connection to earth would cause an RCD or fuse to trip—thus preventing the future electrocution of an occupant handling the appliance. Earthing systems vary between regions, but in countries such as the United Kingdom and Australia both the protective earth and neutral line would be earthed together near the fuse box before the main isolating switch and the neutral earthed once again back at the distribution transformer.

There have been a number of minor changes over the year to practice of residential wiring. Some of the most significant ways modern residential power systems tend to vary from older ones include:

- For convenience, miniature circuit breakers are now almost always used in the fuse box instead of fuses as these can easily be reset by occupants.

- For safety reasons, RCDs are now installed on appliance circuits and, increasingly, even on lighting circuits.

- Dwellings are typically connected to all three-phases of the distribution system with the phases being arbitrarily allocated to the house's single-phase circuits.

- Whereas air conditioners of the past might have been fed from a dedicated circuit attached to a single phase, centralised air conditioners that require three-phase power are now becoming common.

- Protective earths are now run with lighting circuits to allow for metallic lamp holders to be earthed.

- Increasingly residential power systems are incorporating microgenerators, most notably, photovoltaic cells.

Commercial Power Systems

Commercial power systems such as shopping centers or high-rise buildings are larger in scale than residential systems. Electrical designs for larger commercial systems are usually studied for load flow, short-circuit fault levels, and voltage drop for steady-state loads and during starting of large motors. The objectives of the studies are to assure proper equipment and conductor sizing, and to coordinate protective devices so that minimal disruption is cause when a fault is cleared. Large commercial installations will have an orderly system of sub-panels, separate from the main distribution board to allow for better system protection and more efficient electrical installation.

Typically one of the largest appliances connected to a commercial power system is the HVAC unit, and ensuring this unit is adequately supplied is an important consideration in commercial power systems. Regulations for commercial establishments place other requirements on commercial systems that are not placed on residential systems. For example, in Australia, commercial systems must comply with AS 2293, the standard for emergency lighting, which requires emergency lighting be maintained for at least 90 minutes in the event of loss of mains supply. In the United States, the National Electrical Code requires commercial systems to be built with at least one 20A sign outlet in order to light outdoor signage. Building code regulations may place special requirements on the electrical system for emergency lighting, evacuation, emergency power, smoke control and fire protection.

Programmable Logic Controller

Siemens Simatic S7-400 system in a rack, left-to-right: power supply unit (PS), CPU, interface module (IM) and communication processor (CP).

A programmable logic controller, PLC, or programmable controller is a digital computer used for automation of typically industrial electromechanical processes, such as control of machinery on factory assembly lines, amusement rides, or light fixtures. PLCs are used in many machines, in many industries. PLCs are designed for multiple arrangements of digital and analog inputs and outputs, extended temperature ranges, immunity to electrical noise, and resistance to vibration and impact. Programs to control machine operation are typically stored in battery-backed-up or non-volatile memory. A PLC is an example of a "hard" real-time system since output results must be produced in response to input conditions within a limited time, otherwise unintended operation will result.

Before the PLC, control, sequencing, and safety interlock logic for manufacturing automobiles was mainly composed of relays, cam timers, drum sequencers, and dedicated closed-loop controllers. Since these could number in the hundreds or even thousands, the process for updating such facilities for the yearly model change-over was very time consuming and expensive, as electricians needed to individually rewire the relays to change their operational characteristics.

Digital computers, being general-purpose programmable devices, were soon applied to control industrial processes. Early computers required specialist programmers, and stringent operating environmental control for temperature, cleanliness, and power quality. Using a general-purpose computer for process control required protecting the computer from the plant floor conditions. An industrial control computer would have several attributes: it would tolerate the shop-floor environment, it would support discrete (bit-form) input and output in an easily extensible manner, it would not require years of training to use, and it would permit its operation to be monitored. The response time of any computer system must be fast enough to be useful for control; the required speed varying according to the nature of the process. Since many industrial processes have timescales easily addressed by millisecond response times, modern (fast, small, reliable) electronics greatly facilitate building reliable controllers, especially because performance can be traded off for reliability.

In 1968 GM Hydra-Matic (the automatic transmission division of General Motors) issued a request for proposals for an electronic replacement for hard-wired relay systems based on a white paper written by engineer Edward R. Clark. The winning proposal came from Bedford Associates of Bedford, Massachusetts. The first PLC, designated the 084 because it was Bedford Associates' eighty-fourth project, was the result. Bedford Associates started a new company dedicated to developing, manufacturing, selling, and servicing this new product: Modicon, which stood for MOdular DIgital CONtroller. One of the people who worked on that project was Dick Morley, who is considered to be the "father" of the PLC. The Modicon brand was sold in 1977 to Gould Electronics, later acquired by German Company AEG, and then by French Schneider Electric, the current owner.

One of the very first 084 models built is now on display at Modicon's headquarters in North Andover, Massachusetts. It was presented to Modicon by GM, when the unit was retired after nearly twenty years of uninterrupted service. Modicon used the 84 moniker at the end of its product range until the 984 made its appearance.

The automotive industry is still one of the largest users of PLCs.

Early PLCs were designed to replace relay logic systems. These PLCs were programmed in "ladder logic", which strongly resembles a schematic diagram of relay logic. This program notation was chosen to reduce training demands for the existing technicians. Other early PLCs used a form of instruction list programming, based on a stack-based logic solver.

Modern PLCs can be programmed in a variety of ways, from the relay-derived ladder logic to programming languages such as specially adapted dialects of BASIC and C. Another method is state logic, a very high-level programming language designed to program PLCs based on state transition diagrams.

Many early PLCs did not have accompanying programming terminals that were capable of graphical representation of the logic, and so the logic was instead represented as a series of logic expressions in some version of Boolean format, similar to Boolean algebra. As programming terminals evolved, it became more common for ladder logic to be used, for the aforementioned reasons and because it was a familiar format used for electromechanical control panels. Newer formats such as state logic and Function Block (which is similar to the way logic is depicted when using digital integrated logic circuits) exist, but they are still not as popular as ladder logic. A primary reason for this is that PLCs solve the logic in a predictable and repeating sequence, and ladder logic allows the programmer (the person writing the logic) to see any issues with the timing of the logic sequence more easily than would be possible in other formats.

Programming

Early PLCs, up to the mid-1990s, were programmed using proprietary programming panels or special-purpose programming terminals, which often had dedicated function keys representing the various logical elements of PLC programs. Some proprietary programming terminals displayed the elements of PLC programs as graphic symbols, but plain ASCII character representations of contacts, coils, and wires were common. Programs were stored on cassette tape cartridges. Facilities for printing and documentation were minimal due to lack of memory capacity. The oldest PLCs used non-volatile magnetic core memory.

More recently, PLCs are programmed using application software on personal computers, which now represent the logic in graphic form instead of character symbols. The computer is connected to the PLC through Ethernet, RS-232, RS-485, or RS-422 cabling. The programming software allows entry and editing of the ladder-style logic. Generally the software provides functions for debugging and troubleshooting the PLC software, for example, by highlighting portions of the logic to show current status during operation or via simulation. The software will upload and download the PLC program, for backup and restoration purposes. In some models of programmable controller, the program is transferred from a personal computer to the PLC through a programming board which writes the program into a removable chip such as an EPROM

Functionality

The functionality of the PLC has evolved over the years to include sequential relay control, motion control, process control, distributed control systems, and networking. The data handling, storage, processing power, and communication capabilities of some modern PLCs are approximately equivalent to desktop computers. PLC-like programming combined with remote I/O hardware, allow a general-purpose desktop computer to overlap some PLCs in certain applications. Desktop computer controllers have not been generally accepted in heavy industry because the desktop computers run on less stable operating systems than do PLCs, and because the desktop computer hardware is typically not designed to the same levels of tolerance to temperature, humidity, vibration, and longevity as the processors used in PLCs. Operating systems such as Windows do not lend themselves to deterministic logic execution, with the result that the controller may not always respond to changes of input status with the consistency in timing expected from PLCs. Desktop logic applications find use in less critical situations, such as laboratory automation and use in small facilities where the application is less demanding and critical, because they are generally much less expensive than PLCs.

Programmable Logic Relay (PLR)

In more recent years, small products called PLRs (programmable logic relays), and also by similar names, have become more common and accepted. These are much like PLCs, and are used in light industry where only a few points of I/O (i.e. a few signals coming in from the real world and a few going out) are needed, and low cost is desired. These small devices are typically made in a common physical size and shape by several manufacturers, and branded by the makers of larger PLCs to fill out their low end product range. Popular names include PICO Controller, NANO PLC, and other names implying very small controllers. Most of these have 8 to 12 discrete inputs, 4 to 8 discrete outputs, and up to 2 analog inputs. Size is usually about 4" wide, 3" high, and 3" deep. Most such devices include a tiny postage-stamp-sized LCD screen for viewing simplified ladder logic (only a very small portion of the program being visible at a given time) and status of I/O points, and typically these screens are accompanied by a 4-way rocker push-button plus four more separate push-buttons, similar to the key buttons on a VCR remote control, and used to navigate and edit the logic. Most have a small plug for connecting via RS-232 or RS-485 to a personal computer so that programmers can use simple Windows applications for programming instead of being forced to use the tiny LCD and push-button set for this purpose. Unlike regular PLCs that are usually modular and greatly expandable, the PLRs are usually not modular or expandable, but their price can be two orders of magnitude less than a PLC, and they still offer robust design and deterministic execution of the logics.

PLC Topics

Features

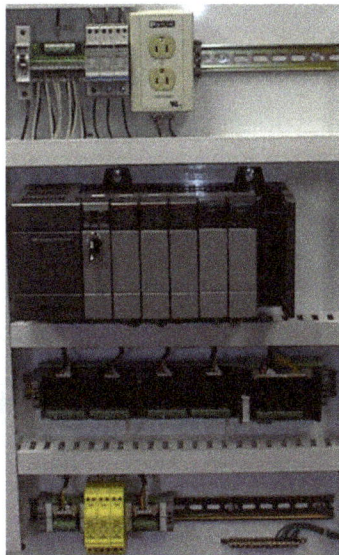

Control panel with PLC (grey elements in the center). The unit consists of separate elements, from left to right; power supply, controller, relay units for in- and output

The main difference from other computers is that PLCs are armored for severe conditions (such as dust, moisture, heat, cold), and have the facility for extensive input/output (I/O) arrangements. These connect the PLC to sensors and actuators. PLCs read limit switches, analog process variables (such as temperature and pressure), and the positions of complex positioning systems. Some use machine vision. On the actuator side, PLCs operate electric motors, pneumatic or hydraulic cylinders, magnetic relays, solenoids, or analog outputs. The input/output arrangements may be built into a simple PLC, or the PLC may have external I/O modules attached to a computer network that plugs into the PLC.

Scan Time

A PLC program is generally executed repeatedly as long as the controlled system is running. The status of physical input points is copied to an area of memory accessible to the processor, sometimes called the "I/O Image Table". The program is then run from its first instruction rung down to the last rung. It takes some time for the processor of the PLC to evaluate all the rungs and update the I/O image table with the status of outputs. This scan time may be a few milliseconds for a small program or on a fast processor, but older PLCs running very large programs could take much longer (say, up to 100 ms) to execute the program. If the scan time were too long, the response of the PLC to process conditions would be too slow to be useful.

As PLCs became more advanced, methods were developed to change the sequence of ladder execution, and subroutines were implemented. This simplified programming could be used to save scan time for high-speed processes; for example, parts of the program used only for setting up the machine could be segregated from those parts required to operate at higher speed.

Special-purpose I/O modules may be used where the scan time of the PLC is too long to allow predictable performance. Precision timing modules, or counter modules for use with shaft encoders, are used where the scan time would be too long to reliably count pulses or detect the sense of rotation of an encoder. The relatively slow PLC can still interpret the counted values to control a machine, but the accumulation of pulses is done by a dedicated module that is unaffected by the speed of the program execution.

System Scale

A small PLC will have a fixed number of connections built in for inputs and outputs. Typically, expansions are available if the base model has insufficient I/O.

Modular PLCs have a chassis (also called a rack) into which are placed modules with different functions. The processor and selection of I/O modules are customized for the particular application. Several racks can be administered by a single processor, and may have thousands of inputs and outputs. Either a special high speed serial I/O link or comparable communication method is used so that racks can be distributed away from the processor, reducing the wiring costs for large plants. Options are also available to mount I/O points directly to the machine and utilize quick disconnecting cables to sensors and valves, saving time for wiring and replacing components.

User Interface

PLCs may need to interact with people for the purpose of configuration, alarm reporting, or everyday control. A human-machine interface (HMI) is employed for this purpose. HMIs are also referred to as man-machine interfaces (MMIs) and graphical user interfaces (GUIs). A simple system may use buttons and lights to interact with the user. Text displays are available as well as graphical touch screens. More complex systems use programming and monitoring software installed on a computer, with the PLC connected via a communication interface.

Communications

PLCs have built-in communications ports, usually 9-pin RS-232, RS-422, RS-485, Ethernet. Various protocols are usually included. Many of these protocols are vendor specific.

Most modern PLCs can communicate over a network to some other system, such as a computer running a SCADA (Supervisory Control And Data Acquisition) system or web browser.

PLCs used in larger I/O systems may have peer-to-peer (P2P) communication between processors. This allows separate parts of a complex process to have individual control while allowing the subsystems to co-ordinate over the communication link. These communication links are also often used for HMI devices such as keypads or PC-type workstations.

Formerly, some manufacturers offered dedicated communication modules as an add-on function where the processor had no network connection built-in.

Programming

PLC programs are typically written in a special application on a personal computer, then downloaded by a direct-connection cable or over a network to the PLC. The program is stored in the PLC either in battery-backed-up RAM or some other non- volatile flash memory. Often, a single PLC can be programmed to replace thousands of relays.

Under the IEC 61131-3 standard, PLCs can be programmed using standards-based programming languages. The most commonly used programming language is Ladder diagram (LD) also known as Ladder logic. It uses Contact-Coil logic to make programs like an electrical control diagram. A graphical programming notation called Sequential Function Charts is available on certain programmable controllers. A model which emulated electromechanical control panel devices (such as the contact and coils of relays) which PLCs replaced. This model remains common today.

IEC 61131-3 currently defines five programming languages for programmable control systems: function block diagram (FBD), ladder diagram (LD), structured text (ST; similar to the Pascal programming language), instruction list (IL; similar to assembly language), and sequential function chart (SFC). These techniques emphasize logical organization of operations.

While the fundamental concepts of PLC programming are common to all manufacturers, differences in I/O addressing, memory organization, and instruction sets mean that PLC programs are never perfectly interchangeable between different makers. Even within the same product line of a single manufacturer, different models may not be directly compatible.

Security

Prior to the discovery of the Stuxnet computer worm in June 2010, security of PLCs received little attention. PLCs generally contain a real-time operating system such as OS-9 or VxWorks, and exploits for these systems exist much as they do for desktop computer operating systems such as Microsoft Windows. PLCs can also be attacked by gaining control of a computer they communicate with.

Simulation

In order to properly understand the operation of a PLC, it is necessary to spend considerable time programming, testing, and debugging PLC programs. PLC systems are inherently expensive, and down-time is often very costly. In addition, if a PLC is programmed incorrectly it can result in lost productivity and dangerous conditions. PLC simulation software such as PLCLogix can save time in the design of automated control applications and can also increase the level of safety associated with equipment since various "what if" scenarios can be tried and tested before the system is activated.

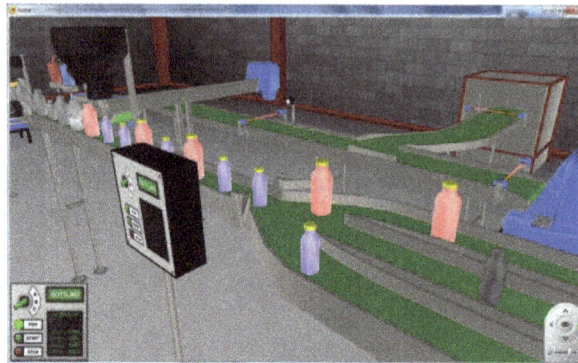

PLCLogix PLC Simulation Software

Redundancy

Some special processes need to work permanently with minimum unwanted down time. Therefore, it is necessary to design a system which is fault-tolerant and capable of handling the process with faulty modules. In such cases to increase the system availability in the event of hardware component failure, redundant CPU or I/O modules with the same functionality can be added to hardware configuration for preventing total or partial process shutdown due to hardware failure.

PLC Compared with Other Control Systems

Allen-Bradley PLC installed in a control panel

PLCs are well adapted to a range of automation tasks. These are typically industrial processes in manufacturing where the cost of developing and maintaining the automation system is high relative to the total cost of the automation, and where changes to the system would be expected during its operational life. PLCs contain input and output devices compatible with industrial pilot devices and controls; little electrical design is required, and the design problem centers on expressing the desired sequence of operations. PLC applications are typically highly customized systems, so the cost of a packaged PLC is low compared to the cost of a specific custom-built controller design. On the other hand, in the case of mass-produced goods, customized control systems are economical. This is due to the lower cost of the components, which can be optimally chosen instead of a "generic" solution, and where the non-recurring engineering charges are spread over thousands or millions of units.

For high volume or very simple fixed automation tasks, different techniques are used. For example, a consumer dishwasher would be controlled by an electromechanical cam timer costing only a few dollars in production quantities.

A microcontroller-based design would be appropriate where hundreds or thousands of units will be produced and so the development cost (design of power supplies, input/output hardware, and necessary testing and certification) can be spread over many sales, and where the end-user would not need to alter the control. Automotive applications are an example; millions of units are built each year, and very few end-users alter the programming of these controllers. However, some specialty vehicles such as transit buses economically use PLCs instead of custom-designed controls, because the volumes are low and the development cost would be uneconomical.

Very complex process control, such as used in the chemical industry, may require algorithms and performance beyond the capability of even high-performance PLCs. Very high-speed or precision controls may also require customized solutions; for example, aircraft flight controls. Single-board computers using semi-customized or fully proprietary hardware may be chosen for very demanding control applications where the high development and maintenance cost can be supported. "Soft PLCs" running on desktop-type computers can interface with industrial I/O hardware while executing programs within a version of commercial operating systems adapted for process control needs.

Programmable controllers are widely used in motion, positioning, and/or torque control. Some manufacturers produce motion control units to be integrated with PLC so that G-code (involving a CNC machine) can be used to instruct machine movements.

PLCs may include logic for single-variable feedback analog control loop, a proportional, integral, derivative (PID) controller. A PID loop could be used to control the temperature of a manufacturing process, for example. Historically PLCs were usually configured with only a few analog control loops; where processes required hundreds or thousands of loops, a distributed control system (DCS) would instead be used. As PLCs have become more powerful, the boundary between DCS and PLC applications has been blurred.

PLCs have similar functionality as remote terminal units. An RTU, however, usually does not support control algorithms or control loops. As hardware rapidly becomes more powerful and cheaper, RTUs, PLCs, and DCSs are increasingly beginning to overlap in responsibilities, and many vendors sell RTUs with PLC-like features, and vice versa. The industry has standardized on the IEC 61131-3 functional block language for creating programs to run on RTUs and PLCs, although nearly all vendors also offer proprietary alternatives and associated development environments.

In recent years "safety" PLCs have started to become popular, either as standalone models or as functionality and safety-rated hardware added to existing controller architectures (Allen-Bradley Guardlogix, Siemens F-series etc.). These differ from conventional PLC types as being suitable for use in safety-critical applications for which PLCs have traditionally been supplemented with hard-wired safety relays. For example, a safety PLC might be used to control access to a robot cell with trapped-key access, or perhaps to manage the shutdown response to an emergency stop on a conveyor production line. Such PLCs typically have a restricted regular instruction set augmented with safety-specific instructions designed to interface with emergency stops, light screens, and so forth. The flexibility that such systems offer has resulted in rapid growth of demand for these controllers.

Discrete and Analog Signals

Discrete signals behave as binary switches, yielding simply an On or Off signal (1 or 0, True or False, respectively). Push buttons, limit switches, and photoelectric sensors are examples of devices providing a discrete signal. Discrete signals are sent using either voltage or current, where a specific range is designated as *On* and another as *Off*. For example, a PLC might use 24 V DC I/O, with values above 22 V DC representing *On*, values below 2VDC representing *Off*, and intermediate values undefined. Initially, PLCs had only discrete I/O.

Analog signals are like volume controls, with a range of values between zero and full-scale. These are typically interpreted as integer values (counts) by the PLC, with various ranges of accuracy depending on the device and the number of bits available to store the data. As PLCs typically use 16-bit signed binary processors, the integer values are limited between -32,768 and +32,767. Pressure, temperature, flow, and weight are often represented by analog signals. Analog signals can use voltage or current with a magnitude proportional to the value of the process signal. For example, an analog 0 to 10 V or 4-20 mA input would be converted into an integer value of 0 to 32767.

Current inputs are less sensitive to electrical noise (e.g. from welders or electric motor starts) than voltage inputs.

Example

As an example, say a facility needs to store water in a tank. The water is drawn from the tank by another system, as needed, and our example system must manage the water level in the tank by controlling the valve that refills the tank. Shown is a "ladder diagram" which shows the control system. A ladder diagram is a method of drawing control circuits which pre-dates PLCs. The ladder diagram resembles the schematic diagram of a system built with electromechanical relays. Shown are:

- Two inputs (from the low and high level switches) represented by contacts of the float switches

- An output to the fill valve, labelled as the fill valve which it controls

- An "internal" contact, representing the output signal to the fill valve which is created in the program.

- A logical control scheme created by the interconnection of these items in software

In ladder diagram, the contact symbols represent the state of bits in processor memory, which corresponds to the state of physical inputs to the system. If a discrete input is energized, the memory bit is a 1, and a "normally open" contact controlled by that bit will pass a logic "true" signal on to the next element of the ladder. Therefore, the contacts in the PLC program that "read" or look at the physical switch contacts in this case must be "opposite" or open in order to return a TRUE for the closed physical switches. Internal status bits, corresponding to the state of discrete outputs, are also available to the program.

In the example, the physical state of the float switch contacts must be considered when choosing "normally open" or "normally closed" symbols in the ladder diagram. The PLC has two discrete inputs from float switches (Low Level and High Level). Both float switches (normally closed) open their contacts when the water level in the tank is above the physical location of the switch.

When the water level is below both switches, the float switch physical contacts are both closed, and a true (logic 1) value is passed to the Fill Valve output. Water begins to fill the tank. The internal "Fill Valve" contact latches the circuit so that even when the "Low Level" contact opens (as the water passes the lower switch), the fill valve remains on. Since the High Level is also normally closed, water continues to flow as the water level remains between the two switch levels. Once the water level rises enough so that the "High Level" switch is off (opened), the PLC will shut the inlet to stop the water from overflowing; this is an example of seal-in (latching) logic. The output is sealed in until a high level condition breaks the circuit. After that the fill valve remains off until the level drops so low that the Low Level switch is activated, and the process repeats again.

A complete program may contain thousands of rungs, evaluated in sequence. Typically the PLC processor will alternately scan all its inputs and update outputs, then evaluate the ladder logic; input changes during a program scan will not be effective until the next I/O update. A complete program scan may take only a few milliseconds, much faster than changes in the controlled process.

Programmable controllers vary in their capabilities for a "rung" of a ladder diagram. Some only allow a single output bit. There are typically limits to the number of series contacts in line, and the number of branches that can be used. Each element of the rung is evaluated sequentially. If elements change their state during evaluation of a rung, hard-to-diagnose faults can be generated, although sometimes (as above) the technique is useful. Some implementations forced evaluation from left-to-right as displayed and did not allow reverse flow of a logic signal (in multi-branched rungs) to affect the output.

PLCs are at the forefront of manufacturing automation. An engineer working in a manufacturing environment will at least encounter some PLCs, if not use them on a regular basis. Electrical engineering students should have basic knowledge of PLCs because of their widespread use in industrial applications.

Root Locus

In control theory and stability theory, root locus analysis is a graphical method for examining how the roots of a system change with variation of a certain system parameter, commonly a gain within a feedback system. This is a technique used as a stability criterion in the field of classical control theory developed by Walter R. Evans which can determine stability of the system. The root locus plots the poles of the closed loop transfer function in the complex s-plane as a function of a gain parameter.

Uses

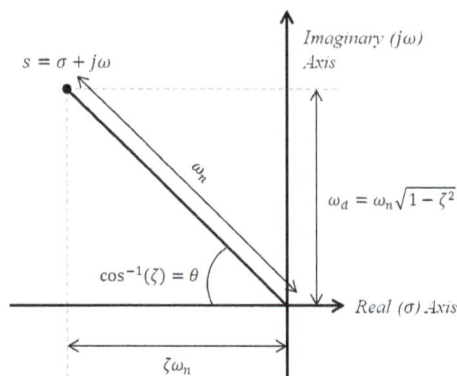

Effect of pole location on a second order system's natural frequency and damping ratio.

In addition to determining the stability of the system, the root locus can be used to design the damping ratio (ζ) and natural frequency (ω_n) of a feedback system. Lines of constant damping ratio can be drawn radially from the origin and lines of constant natural frequency can be drawn as arcs whose center points coincide with the origin. By selecting a point along the root locus that coincides with a desired damping ratio and natural frequency, a gain K can be calculated and implemented in the controller. More elaborate techniques of controller design using the root locus are available in most control textbooks: for instance, lag, lead, PI, PD and PID controllers can be designed approximately with this technique.

The definition of the damping ratio and natural frequency presumes that the overall feedback system is well approximated by a second order system; i.e. the system has a dominant pair of poles. This is often not the case, so it is good practice to simulate the final design to check if the project goals are satisfied.

Definition

The root locus of a feedback system is the graphical representation in the complex s-plane of the possible locations of its closed-loop poles for varying values of a certain system parameter. The points that are part of the root locus satisfy the angle condition. The value of the parameter for a certain point of the root locus can be obtained using the magnitude condition.

Suppose there is a feedback system with input signal $X(s)$ and output signal $Y(s)$. The forward path transfer function is $G(s)$; the feedback path transfer function is $H(s)$.

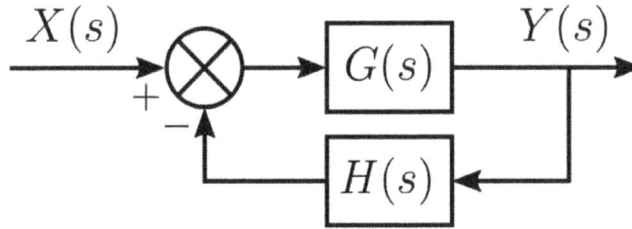

For this system, the closed-loop transfer function is given by

$$T(s) = \frac{Y(s)}{X(s)} = \frac{G(s)}{1+G(s)H(s)}$$

Thus, the closed-loop poles of the closed-loop transfer function are the roots of the characteristic equation $1+G(s)H(s)=0$. The roots of this equation may be found wherever $G(s)H(s)=-1$..

In systems without pure delay, the product $G(s)H(s)$ is a rational polynomial function and may be expressed as

$$G(s)H(s) = K\frac{(s+z_1)(s+z_2)\cdots(s+z_m)}{(s+p_1)(s+p_2)\cdots(s+p_n)}$$

where $-z_i$ are the m zeros, $-p_i$ are the n poles, and K is a scalar gain. Typically, a root locus diagram will indicate the transfer function's pole locations for varying values of the parameter K. . A root locus plot will be all those points in the s-plane where $G(s)H(s)=-1$ for any value of K .

The factoring of K and the use of simple monomials means the evaluation of the rational polynomial can be done with vector techniques that add or subtract angles and multiply or divide magnitudes. The vector formulation arises from the fact that each monomial term $(s-a)$ in the factored $G(s)H(s)$ represents the vector from a to s in the s-plane. The polynomial can be evaluated by considering the magnitudes and angles of each of these vectors.

According to vector mathematics, the angle of the result of the rational polynomial is the sum of all the angles in the numerator minus the sum of all the angles in the denominator. So to test whether a point in the s-plane is on the root locus, only the angles to all the open loop poles and zeros need be considered. This is known as the angle condition.

Similarly, the magnitude of the result of the rational polynomial is the product of all the magnitudes in the numerator divided by the product of all the magnitudes in the denominator. It turns out that the calculation of the magnitude is not needed to determine if a point in the s-plane is part of the root locus because K varies and can take an arbitrary real value. For each point of the root locus a value of K can be calculated. This is known as the magnitude condition.

A graphical method that uses a special protractor called a "Spirule" was once used to determine angles and draw the root loci.

The root locus only gives the location of closed loop poles as the gain K is varied. The value of K does not affect the location of the zeros. The open-loop zeros are the same as the closed-loop zeros.

Angle Condition

A point s of the complex s-plane satisfies the angle condition if

$$s\angle(G(s)H(s)) = \pi$$

which is the same as saying that

$$\sum_{i=1}^{m} \angle(s + z_i) - \sum_{i=1}^{n} \angle(s + p_i) = \pi$$

that is, the sum of the angles from the open-loop zeros to the point s minus the angles from the open-loop poles to the point s has to be equal to π, or 180 degrees.

Magnitude condition

A value of K satisfies the magnitude condition for a given s point of the root locus if

$$|G(s)H(s)| = 1$$

which is the same as saying that

$$K \frac{|s + z_1| |s + z_2| \cdots |s + z_m|}{|s + p_1| |s + p_2| \cdots |s + p_n|} = 1.$$

Sketching Root Locus

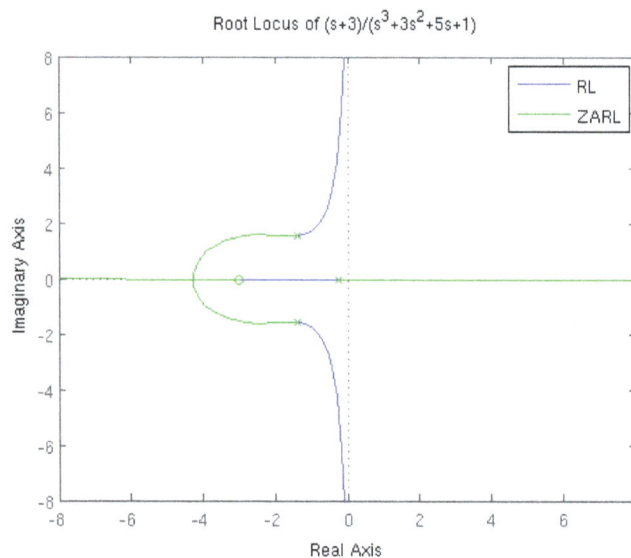

Root Locus of (s+3)/(s³+3s²+5s+1)

RL = root locus; ZARL = zero angle root locus

Using a few basic rules, the root locus method can plot the overall shape of the path (locus) traversed by the roots as the value of K varies. The plot of the root locus then gives an idea of the stability and dynamics of this feedback system for different values of K. The rules are the following:

- Mark open-loop poles and zeros

- Mark real axis portion to the left of an odd number of poles and zeros

- Find asymptotes

Let P be the number of poles and Z be the number of zeros:

$P - Z$ = number of asymptotes

The asymptotes intersect the real axis at α (which is called the centroid) and depart at angle ϕ given by:

$$\phi_l = \frac{180° + (l-1)360°}{P - Z}, l = 1, 2, \ldots, P - Z$$

where is the sum of all the locations of the poles, and is the sum of all the locations of the explicit zeros.

- Phase condition on test point to find angle of departure

- Compute breakaway/break-in points

The breakaway points are located at the roots of the following equation:

$$\frac{dG(s)H(s)}{ds} = 0 \text{ or } \frac{d\overline{GH}(z)}{dz} = 0$$

Once you solve for z, the real roots give you the breakaway/reentry points. Complex roots correspond to a lack of breakaway/reentry.

Z-plane Versus S-plane

The root locus method can also be used for the analysis of sampled data systems by computing the root locus in the z-plane, the discrete counterpart of the s-plane. The equation $z = e^{sT}$ maps continuous s-plane poles (not zeros) into the z-domain, where T is the sampling period. The stable, left half s-plane maps into the interior of the unit circle of the z-plane, with the s-plane origin equating to $|z| = 1$ (because $e^0 = 1$). A diagonal line of constant damping in the s-plane maps around a spiral from (1,0) in the z plane as it curves in toward the origin. Note also that the Nyquist aliasing criteria is expressed graphically in the z-plane by the x-axis, where $\omega nT = \pi$. The line of constant damping just described spirals in indefinitely but in sampled data systems, frequency content is aliased down to lower frequencies by integral multiples of the Nyquist frequency. That is, the sampled response appears as a lower frequency and better damped as well since the root in the z-plane maps equally well to the first loop of a different, better damped spiral curve of constant damping. Many other interesting and relevant mapping properties can be described, not least that z-plane controllers, having the property that they may be directly implemented from the z-plane transfer function (zero/pole ratio of polynomials), can be imagined graphically on a z-plane plot of the open loop transfer function, and immediately analyzed utilizing root locus.

Since root locus is a graphical angle technique, root locus rules work the same in the z and s planes.

The idea of a root locus can be applied to many systems where a single parameter K is varied. For example, it is useful to sweep any system parameter for which the exact value is uncertain in order to determine its behavior.

Frequency Response

Frequency response is the quantitative measure of the output spectrum of a system or device in response to a stimulus, and is used to characterize the dynamics of the system. It is a measure of magnitude and phase of the output as a function of frequency, in comparison to the input. In simplest terms, if a sine wave is injected into a system at a given frequency, a linear system will respond at that same frequency with a certain magnitude and a certain phase angle relative to the input. Also for a linear system, doubling the amplitude of the input will double the amplitude of the output. In addition, if the system is time-invariant (so LTI), then the frequency response also will not vary with time. Thus for LTI systems, the frequency response can be seen as applying the system's transfer function to a purely imaginary number argument representing the frequency of the sinusoidal excitation.

Two applications of frequency response analysis are related but have different objectives. For an audio system, the objective may be to reproduce the input signal with no distortion. That would require a uniform (flat) magnitude of response up to the bandwidth limitation of the system, with the signal delayed by precisely the same amount of time at all frequencies. That amount of time could be seconds, or weeks or months in the case of recorded media. In contrast, for a feedback apparatus used to control a dynamic system, the objective is to give the closed-loop system improved response as compared to the uncompensated system. The feedback generally needs to respond to system dynamics within a very small number of cycles of oscillation (usually less than one full cycle), and with a definite phase angle relative to the commanded control input. For feedback of sufficient amplification, getting the phase angle wrong can lead to instability for an open-loop stable system, or failure to stabilize a system that is open-loop unstable. Digital filters may be used for both audio systems and feedback control systems, but since the objectives are different, generally the phase characteristics of the filters will be significantly different for the two applications.

Estimation and Plotting

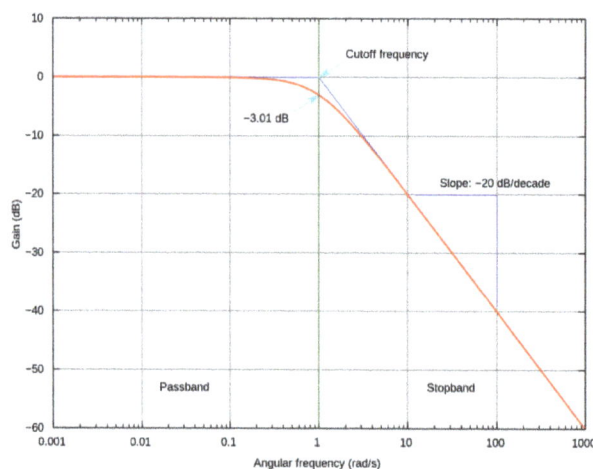

Frequency response of a low pass filter with 6 dB per octave or 20 dB per decade

Estimating the frequency response for a physical system generally involves exciting the system with an input signal, measuring both input and output time histories, and comparing the two through a process such as the Fast Fourier Transform (FFT). One thing to keep in mind for the analysis is that the frequency content of the input signal must cover the frequency range of interest or the results will not be valid for the portion of the frequency range not covered.

The frequency response of a system can be measured by applying a *test signal*, for example:

- applying an impulse to the system and measuring its response

- sweeping a constant-amplitude pure tone through the bandwidth of interest and measuring the output level and phase shift relative to the input

- applying a signal with a wide frequency spectrum (for example digitally-generated maximum length sequence noise, or analog filtered white noise equivalent, like pink noise), and calculating the impulse response by deconvolution of this input signal and the output signal of the system.

The frequency response is characterized by the *magnitude* of the system's response, typically measured in decibels (dB) or as a decimal, and the *phase*, measured in radians or degrees, versus frequency in radians/sec or Hertz (Hz).

These response measurements can be plotted in three ways: by plotting the magnitude and phase measurements on two rectangular plots as functions of frequency to obtain a Bode plot; by plotting the magnitude and phase angle on a single polar plot with frequency as a parameter to obtain a Nyquist plot; or by plotting magnitude and phase on a single rectangular plot with frequency as a parameter to obtain a Nichols plot.

For audio systems with nearly uniform time delay at all frequencies, the magnitude versus frequency portion of the Bode plot may be all that is of interest. For design of control systems, any of the three types of plots [Bode, Nyquist, Nichols] can be used to infer closed-loop stability and stability margins (gain and phase margins) from the open-loop frequency response, provided that for the Bode analysis the phase-versus-frequency plot is included.

Nonlinear Frequency Response

If the system under investigation is nonlinear then applying purely linear frequency domain analysis will not reveal all the nonlinear characteristics. To overcome these limitations, generalized frequency response functions and nonlinear output frequency response functions have been defined that allow the user to analyze complex nonlinear dynamic effects. The nonlinear frequency response methods reveal complex resonance, inter modulation, and energy transfer effects that cannot be seen using a purely linear analysis and are becoming increasingly important in a nonlinear world.

Applications

In electronics this stimulus would be an input signal. In the audible range it is usually referred to in connection with electronic amplifiers, microphones and loudspeakers. Radio spectrum frequency response can refer to measurements of coaxial cable, twisted-pair cable, video switching equipment, wireless communications devices, and antenna systems. Infrasonic frequency response measurements include earthquakes and electroencephalography (brain waves).

Frequency response requirements differ depending on the application. In high fidelity audio, an amplifier requires a frequency response of at least 20–20,000 Hz, with a tolerance as tight as ±0.1 dB in the mid-range frequencies around 1000 Hz, however, in telephony, a frequency response of 400–4,000 Hz, with a tolerance of ±1 dB is sufficient for intelligibility of speech.

Frequency response curves are often used to indicate the accuracy of electronic components or systems. When a system or component reproduces all desired input signals with no emphasis or attenuation of a particular frequency band, the system or component is said to be "flat", or to have a flat frequency response curve.

Once a frequency response has been measured (e.g., as an impulse response), provided the system is linear and time-invariant, its characteristic can be approximated with arbitrary accuracy by a digital filter. Similarly, if a system is demonstrated to have a poor frequency response, a digital or analog filter can be applied to the signals prior to their reproduction to compensate for these deficiencies.

References

- Basso, Christophe (2012). "Designing Control Loops for Linear and Switching Power Supplies: A Tutorial Guide". Artech House, ISBN 978-1608075577

- To cite a book with a credited author Winer, Ethan (2013). "Part 3". The Audio Expert. New York and London: Focal Press. ISBN 978-0-240-82100-9.

- Bela G. Liptak (29 September 2005). Instrument Engineers' Handbook, Fourth Edition, Volume Two: Process Control and Optimization. CRC Press. pp. 518–. ISBN 978-1-4200-6400-1.

- Aquino-Santos, Raul (30 November 2010). Emerging Technologies in Wireless Ad-hoc Networks: Applications and Future Development: Applications and Future Development. IGI Global. pp. 43–. ISBN 978-1-60960-029-7.

- Boyer, Stuart A. (2010). SCADA Supervisory Control and Data Acquisition. USA: ISA - International Society of Automation. p. 179. ISBN 978-1-936007-09-7.

- Slay, J.; Miller, M. (November 2007). "Chpt 6: Lessons Learned from the Maroochy Water Breach". Critical infrastructure protection (Online-Ausg. ed.). Springer Boston. pp. 73–82. ISBN 978-0-387-75461-1. Retrieved 2 May 2012.

- Ned Mohan; T. M. Undeland; William P. Robbins (2003). Power Electronics: Converters, Applications, and Design. United States of America: John Wiley & Sons, Inc. ISBN 0-471-22693-9.

- "Air-conditioner Manufacturer Chooses Smart Power Modules". Power Electronics Technology. 31 August 2005. Retrieved 30 March 2016.

- "ICSA-11-231-01—INDUCTIVE AUTOMATION IGNITION INFORMATION DISCLOSURE VULNERABILITY" (PDF). 19 Aug 2011. Retrieved 21 Jan 2013.

- "ICSA-11-094-01—WONDERWARE INBATCH CLIENT ACTIVEX BUFFER OVERFLOW" (PDF). 13 Apr 2011. Retrieved 26 Mar 2013.

Control Engineering: An Integrated Study

The application of control theory to design systems with desired behavior is certified as control engineering. This chapter is a compilation of the important topics related to control engineering, such as senor and acutuator. It offers an insightful focus, keeping in mind the complex subject matter.

Control Engineering

Control engineering or control systems engineering is the engineering discipline that applies control theory to design systems with desired behaviors. The practice uses sensors to measure the output performance of the device being controlled and those measurements can be used to give feedback to the input actuators that can make corrections toward desired performance. When a device is designed to perform without the need of human inputs for correction it is called automatic control (such as cruise control for regulating the speed of a car). Multi-disciplinary in nature, control systems engineering activities focus on implementation of control systems mainly derived by mathematical modeling of systems of a diverse range.

Control systems play a critical role in space flight

Overview

Modern day control engineering is a relatively new field of study that gained significant attention during the 20th century with the advancement of technology. It can be broadly defined or classified as practical application of control theory. Control engineering has an essential role in a wide range of control systems, from simple household washing machines to high-performance F-16 fighter aircraft. It seeks to understand physical systems, using mathematical modeling, in terms of inputs, outputs and various components with different behaviors, use control systems design tools to develop controllers for those systems and implement controllers in physical systems employing available technology. A system can be mechanical, electrical, fluid, chemical, financial and even biological, and the mathematical modeling, analysis and controller design uses control theory in one or many of the time, frequency and complex-s domains, depending on the nature of the design problem.

History

Automatic control systems were first developed over two thousand years ago. The first feedback control device on record is thought to be the ancient Ktesibios's water clock in Alexandria, Egypt around the third century B.C. It kept time by regulating the water level in a vessel and, therefore, the water flow from that vessel. This certainly was a successful device as water clocks of similar design were still being made in Baghdad when the Mongols captured the city in 1258 A.D. A variety of automatic devices have been used over the centuries to accomplish useful tasks or simply to just entertain. The latter includes the automata, popular in Europe in the 17th and 18th centuries, featuring dancing figures that would repeat the same task over and over again; these automata are examples of open-loop control. Milestones among feedback, or "closed-loop" automatic control devices, include the temperature regulator of a furnace attributed to Drebbel, circa 1620, and the centrifugal flyball governor used for regulating the speed of steam engines by James Watt in 1788.

In his 1868 paper "On Governors", James Clerk Maxwell was able to explain instabilities exhibited by the flyball governor using differential equations to describe the control system. This demonstrated the importance and usefulness of mathematical models and methods in understanding complex phenomena, and signaled the beginning of mathematical control and systems theory. Elements of control theory had appeared earlier but not as dramatically and convincingly as in Maxwell's analysis.

Control theory made significant strides in the next 100 years. New mathematical techniques made it possible to control, more accurately, significantly more complex dynamical systems than the original flyball governor. These techniques include developments in optimal control in the 1950s and 1960s, followed by progress in stochastic, robust, adaptive and optimal control methods in the 1970s and 1980s. Applications of control methodology have helped make possible space travel and communication satellites, safer and more efficient aircraft, cleaner auto engines, cleaner and more efficient chemical processes.

Before it emerged as a unique discipline, control engineering was practiced as a part of mechanical engineering and control theory was studied as a part of electrical engineering since electrical circuits can often be easily described using control theory techniques. In the very first control relationships, a current output was represented with a voltage control input. However, not having proper technology to implement electrical control systems, designers left with the option of less efficient and slow responding mechanical systems. A very effective mechanical controller that is still widely used in some hydro plants is the governor. Later on, previous to modern power electronics, process control systems for industrial applications were devised by mechanical engineers using pneumatic and hydraulic control devices, many of which are still in use today.

Control Theory

There are two major divisions in control theory, namely, classical and modern, which have direct implications over the control engineering applications. The scope of classical control theory is limited to single-input and single-output (SISO) system design, except when analyzing for disturbance rejection using a second input. The system analysis is carried out in the time domain using differential equations, in the complex-s domain with the Laplace transform, or in the frequency domain by transforming from the complex-s domain. Many systems may be assumed to have a second order and single variable system response in the time domain. A controller designed using classical theory often requires on-site tuning due to incorrect design approximations. Yet, due to the easier physical implementation of classical controller designs as compared to systems designed using modern control theory, these controllers are preferred in most industrial applications. The most common controllers designed using classical control theory are PID controllers. A less common implementation may include either or both a Lead or Lag filter. The ultimate end goal is to meet a requirements set typically provided in the time-domain called the Step response, or at times in the frequency domain called the Open-Loop response. The Step response characteristics applied in a specification are typically percent overshoot, settling time, etc. The Open-Loop response characteristics applied in a specification are typically Gain and Phase margin and bandwidth. These characteristics may be evaluated through simulation including a dynamic model of the system under control coupled with the compensation model.

In contrast, modern control theory is carried out in the state space, and can deal with multiple-input and multiple-output (MIMO) systems. This overcomes the limitations of classical control theory in more sophisticated design problems, such as fighter aircraft control, with the limitation that no frequency domain analysis is possible. In modern design, a system is represented to the greatest advantage as a set of decoupled first order differential equations defined using state variables. Nonlinear, multivariable, adaptive and robust control theories come under this division. Matrix methods are significantly limited for MIMO systems where linear independence cannot be assured in the relationship between inputs and outputs. Being fairly new, modern control theory has many areas yet to be explored. Scholars like Rudolf E. Kalman and Aleksandr Lyapunov are well-known among the people who have shaped modern control theory.

Control Systems

Control engineering is the engineering discipline that focuses on the modeling of a diverse range of dynamic systems (e.g. mechanical systems) and the design of controllers that will cause these systems to behave in the desired manner. Although such controllers need not be electrical many are and hence control engineering is often viewed as a subfield of electrical engineering. However, the falling price of microprocessors is making the actual implementation of a control system essentially trivial. As a result, focus is shifting back to the mechanical and process engineering discipline, as intimate knowledge of the physical system being controlled is often desired.

Electrical circuits, digital signal processors and microcontrollers can all be used to implement control systems. Control engineering has a wide range of applications from the flight and propulsion systems of commercial airliners to the cruise control present in many modern automobiles.

In most of the cases, control engineers utilize feedback when designing control systems. This is often accomplished using a PID controller system. For example, in an automobile with cruise control the vehicle's speed is continuously monitored and fed back to the system, which adjusts the motor's torque accordingly. Where there is regular feedback, control theory can be used to determine how the system responds to such feedback. In practically all such systems stability is important and control theory can help ensure stability is achieved.

Although feedback is an important aspect of control engineering, control engineers may also work on the control of systems without feedback. This is known as open loop control. A classic example of open loop control is a washing machine that runs through a pre-determined cycle without the use of sensors.

Control Engineering Education

At many universities, control engineering courses are taught in electrical and electronic engineering, mechatronics engineering, mechanical engineering, and aerospace engineering. In others, control engineering is connected to computer science, as most control techniques today are implemented through computers, often as embedded systems (as in the automotive field). The field of control within chemical engineering is often known as process control. It deals primarily with the control of variables in a chemical process in a plant. It is taught as part of the undergraduate curriculum of any chemical engineering program and employs many of the same principles in control engineering. Other engineering disciplines also overlap with control engineering as it can be applied to any system for which a suitable model can be derived. However, specialised control engineering departments do exist, for example, the Department of Automatic Control and Systems Engineering at the University of Sheffield and the Department of Systems Engineering at the United States Naval Academy.

Control engineering has diversified applications that include science, finance management, and even human behavior. Students of control engineering may start with a linear control system course dealing with the time and complex-s domain, which requires a thorough background in elementary mathematics and Laplace transform, called classical control theory. In linear control, the student does frequency and time domain analysis. Digital control and nonlinear control courses require Z transformation and algebra respectively, and could be said to complete a basic control education.

Recent advancement

Originally, control engineering was all about continuous systems. Development of computer control tools posed a requirement of discrete control system engineering because the communications between the computer-based digital controller and the physical system are governed by a computer clock. The equivalent to Laplace transform in the discrete domain is the Z-transform. Today, many of the control systems are computer controlled and they consist of both digital and analog components.

Therefore, at the design stage either digital components are mapped into the continuous domain and the design is carried out in the continuous domain, or analog components are mapped into discrete domain and design is carried out there. The first of these two methods is more commonly encountered in practice because many industrial systems have many continuous systems components, including mechanical, fluid, biological and analog electrical components, with a few digital controllers.

Similarly, the design technique has progressed from paper-and-ruler based manual design to computer-aided design and now to computer-automated design or CAutoD which has been made possible by evolutionary computation. CAutoD can be applied not just to tuning a predefined control scheme, but also to controller structure optimisation, system identification and invention of novel control systems, based purely upon a performance requirement, independent of any specific control scheme.

Resilient Control Systems extends the traditional focus on addressing only plant disturbances to frameworks, architectures and methods that address multiple types of unexpected disturbance. In particular, adapting and transforming behaviors of the control system in response to malicious actors, abnormal failure modes, undesirable human action, etc. Development of resilience technologies require the involvement of multidisciplinary teams to holistically address the performance challenges.

Sensor

In the broadest definition, a sensor is an object whose purpose is to detect events or changes in its environment, and then provide a corresponding output. A sensor is a type of transducer; sensors may provide various types of output, but typically use electrical or optical signals. For example, a thermocouple generates a known voltage (the output) in response to its temperature (the environment). A mercury-in-glass thermometer, similarly, converts measured temperature into expansion and contraction of a liquid, which can be read on a calibrated glass tube.

Sensors are used in everyday objects such as touch-sensitive elevator buttons (tactile sensor) and lamps which dim or brighten by touching the base, besides innumerable applications of which most people are never aware. With advances in micromachinery and easy-to-use micro controller platforms, the uses of sensors have expanded beyond the most traditional fields of temperature, pressure or flow measurement, for example into MARG sensors. Moreover, analog sensors such as potentiometers and force-sensing resistors are still widely used. Applications include manufacturing and machinery, airplanes and aerospace, cars, medicine, and robotics.it is also included in our day-to-day life.

A sensor's sensitivity indicates how much the sensor's output changes when the input quantity being measured changes. For instance, if the mercury in a thermometer moves 1 cm when the temperature changes by 1 °C, the sensitivity is 1 cm/°C (it is basically the slope Dy/Dx assuming a linear characteristic). Some sensors can also affect what they measure; for instance, a room temperature thermometer inserted into a hot cup of liquid cools the liquid while the liquid heats the thermometer. Sensors need to be designed to have a small effect on what is measured; making the sensor smaller often improves this and may introduce other advantages. Technological progress allows more and more sensors to be manufactured on a microscopic scale as microsensors using MEMS technology. In most cases, a microsensor reaches a significantly higher speed and sensitivity compared with macroscopic approaches.

Classification of Measurement Errors

A good sensor obeys the following rules::

- it is sensitive to the measured property,

- it is insensitive to any other property likely to be encountered in its application, and

- it does not influence the measured property.

The sensitivity is then defined as the ratio between the output signal and measured property. For example, if a sensor measures temperature and has a voltage output, the sensitivity is a constant with the unit [V/K]; this sensor is linear because the ratio is constant at all points of measurement.

For an analog sensor signal to be processed, or used in digital equipment, it needs to be converted to a digital signal, using an analog-to-digital converter.

IR SENSOR (TRANSCEIVER)

An infrared sensor

Sensor Deviations

If the sensor is not ideal, several types of deviations can be observed:

- The sensitivity may in practice differ from the value specified. This is called a sensitivity error.

- Since the range of the output signal is always limited, the output signal will eventually reach a minimum or maximum when the measured property exceeds the limits. The full scale range defines the maximum and minimum values of the measured property.

- If the output signal is not zero when the measured property is zero, the sensor has an offset or bias. This is defined as the output of the sensor at zero input.

- If the sensitivity is not constant over the range of the sensor, this is called nonlinearity. Usually, this is defined by the amount the output differs from ideal behavior over the full range of the sensor, often noted as a percentage of the full range.

- If the deviation is caused by a rapid change of the measured property over time, there is a dynamic error. Often, this behavior is described with a bode plot showing sensitivity error and phase shift as a function of the frequency of a periodic input signal.

- If the output signal slowly changes independent of the measured property, this is defined as drift (telecommunication). Long term drift usually indicates a slow degradation of sensor properties over a long period of time.

- Noise is a random deviation of the signal that varies in time.

- Hysteresis is an error caused by when the measured property reverses direction, but there is some finite lag in time for the sensor to respond, creating a different offset error in one direction than in the other.

- If the sensor has a digital output, the output is essentially an approximation of the measured property. The approximation error is also called digitization error.

- If the signal is monitored digitally, limitation of the sampling frequency also can cause a dynamic error, or if the variable or added noise changes periodically at a frequency near a multiple of the sampling rate may induce aliasing errors.

- The sensor may to some extent be sensitive to properties other than the property being measured. For example, most sensors are influenced by the temperature of their environment.

All these deviations can be classified as systematic errors or random errors. Systematic errors can sometimes be compensated for by means of some kind of calibration strategy. Noise is a random error that can be reduced by signal processing, such as filtering, usually at the expense of the dynamic behavior of the sensor.

Resolution

The resolution of a sensor is the smallest change it can detect in the quantity that it is measuring. Often in a digital display, the least significant digit will fluctuate, indicating that changes of that magnitude are only just resolved. The resolution is related to the precision with which the measurement is made. For example, a scanning tunneling probe (a fine tip near a surface collects an electron tunneling current) can resolve atoms and molecules.

Sensors in Nature

All living organisms contain biological sensors with functions similar to those of the mechanical devices described. Most of these are specialized cells that are sensitive to:

- Light, motion, temperature, magnetic fields, gravity, humidity, moisture, vibration, pressure, electrical fields, sound, and other physical aspects of the external environment

- Physical aspects of the internal environment, such as stretch, motion of the organism, and position of appendages (proprioception)

- Environmental molecules, including toxins, nutrients, and pheromones

- Estimation of biomolecules interaction and some kinetics parameters

- Internal metabolic indicators, such as glucose level, oxygen level, or osmolality

- Internal signal molecules, such as hormones, neurotransmitters, and cytokines

- Differences between proteins of the organism itself and of the environment or alien creatures.

Chemical sensor

A chemical sensor is a self-contained analytical device that can provide information about the chemical composition of its environment, that is, a liquid or a gas phase. The information is provided in the form of a measurable physical signal that is correlated with the concentration of a certain chemical species (termed as analyte). Two main steps are involved in the functioning of a chemical sensor, namely, recognition and transduction. In the recognition step, analyte molecules interact selectively with receptor molecules or sites included in the structure of the recognition element of the sensor. Consequently, a characteristic physical parameter varies and this variation is reported by means of an integrated transducer that generates the output signal. A chemical sensor based on recognition material of biological nature is a biosensor. However, as synthetic biomimetic materials are going to substitute to some extent recognition biomaterials, a sharp distinction between a biosensor and a standard chemical sensor is superfluous. Typical biomimetic materials used in sensor development are molecularly imprinted polymers and aptamers.

Biosensor

In biomedicine and biotechnology, sensors which detect analytes thanks to a biological component, such as cells, protein, nucleic acid or biomimetic polymers, are called biosensors. Whereas a non-biological sensor, even organic (=carbon chemistry), for biological analytes is referred to as sensor or nanosensor. This terminology applies for both in-vitro and in vivo applications. The encapsulation of the biological component in biosensors, presents a slightly different problem that ordinary sensors; this can either be done by means of a semipermeable barrier, such as a dialysis membrane or a hydrogel, or a 3D polymer matrix, which either physically constrains the sensing macromolecule or chemically constrains the macromolecule by bounding it to the scaffold.

Actuator

An actuator is a component of machines that is responsible for moving or controlling a mechanism or system.

An actuator requires a control signal and source of energy. The control signal is relatively low energy and may be electric voltage or current, pneumatic or hydraulic pressure, or even human power. The supplied main energy source may be electric current, hydraulic fluid pressure, or pneumatic pressure. When the control signal is received, the actuator responds by converting the energy into mechanical motion.

An actuator is the mechanism by which a control system acts upon an environment. The control system can be simple (a fixed mechanical or electronic system), software-based (e.g. a printer driver, robot control system), a human, or any other input.

History

The history of the pneumatic actuation system and the hydraulic actuation system dates to around the time of World War II (1938). It was first created by Xhiter Anckeleman (pronounced 'Ziter') who used his knowledge of engines and brake systems to come up with a new solution to ensure that the brakes on a car exert the maximum force, with the least possible wear and tear.

Hydraulic

A hydraulic actuator consists of cylinder or fluid motor that uses hydraulic power to facilitate mechanical operation. The mechanical motion gives an output in terms of linear, rotary or oscillatory motion. Because liquids are nearly impossible to compress, a hydraulic actuator can exert a large force. The drawback of this approach is its limited acceleration.

The hydraulic cylinder consists of a hollow cylindrical tube along which a piston can slide. The term *single acting* is used when the fluid pressure is applied to just one side of the piston. The piston can move in only one direction, a spring being frequently used to give the piston a return stroke. The term *double acting* is used when pressure is applied on each side of the piston; any difference in pressure between the two side of the piston moves the piston to one side or the other.

Pneumatic

Pneumatic rack and pinion actuators for valve controls of water pipes

A pneumatic actuator converts energy formed by vacuum or compressed air at high pressure into either linear or rotary motion. Pneumatic energy is desirable for main engine controls because it can quickly respond in starting and stopping as the power source does not need to be stored in reserve for operation.

Pneumatic actuators enable considerable forces to be produced from relatively small pressure changes. These forces are often used with valves to move diaphragms to affect the flow of liquid through the valve.

Electric

An electric actuator is powered by a motor that converts electrical energy into mechanical torque. The electrical energy is used to actuate equipment such as multi-turn valves. It is one of the cleanest and most readily available forms of actuator because it does not involve oil.

Thermal or Magnetic (Shape Memory Alloys)

Actuators which can be actuated by applying thermal or magnetic energy have been used in commercial applications. They tend to be compact, lightweight, economical and with high power density. These actuators use shape memory materials (SMMs), such as shape memory alloys (SMAs) or magnetic shape-memory alloys (MSMAs). Some popular manufacturers of these devices are Finnish Modti Inc., American Dynalloy and Rotork.

Mechanical

A mechanical actuator functions to execute movement by converting one kind of motion, such as rotary motion, into another kind, such as linear motion. An example is a rack and pinion. The operation of mechanical actuators is based on combinations of structural components, such as gears and rails, or pulleys and chains.

Examples and applications

In engineering, actuators are frequently used as mechanisms to introduce motion, or to clamp an object so as to prevent motion. In electronic engineering, actuators are a subdivision of transducers. They are devices which transform an input signal (mainly an electrical signal) into some form of motion.

Examples of actuators

- Comb drive
- Digital micromirror device
- Electric motor
- Electroactive polymer
- Hydraulic cylinder
- Piezoelectric actuator
- Pneumatic actuator
- Screw jack
- Servomechanism
- Stepper motor
- Shape-memory alloy
- Thermal bimorph

Circular to Linear Conversion

Motors are mostly used when circular motions are needed, but can also be used for linear applications by transforming circular to linear motion with a lead screw or similar mechanism. On the other hand, some actuators are intrinsically linear, such as piezoelectric actuators. Conversion between circular and linear motion is commonly made via a few simple types of mechanism including:

- Screw: Screw jack, ball screw and roller screw actuators all operate on the principle of the simple machine known as the screw. By rotating the actuator's nut, the screw shaft moves in a line. By moving the screw shaft, the nut rotates.

- Wheel and axle: Hoist, winch, rack and pinion, chain drive, belt drive, rigid chain and rigid belt actuators operate on the principle of the wheel and axle. By rotating a wheel/axle (e.g. drum, gear, pulley or shaft) a linear member (e.g. cable, rack, chain or belt) moves. By moving the linear member, the wheel/axle rotates.

Virtual Instrumentation

In virtual instrumentation, actuators and sensors are the hardware complements of virtual instruments.

Performance metrics

Performance metrics for actuators include speed, acceleration, and force (alternatively, angular speed, angular acceleration, and torque), as well as energy efficiency and considerations such as mass, volume, operating conditions, and durability, among others.

Force

When considering force in actuators for applications, two main metrics should be considered. These two are static and dynamic loads. Static load is the force capability of the actuator while not in motion. Conversely, the dynamic load of the actuator is the force capability while in motion. The two aspects rarely have the same weight capability and must be considered separately.

Speed

Speed should be considered primarily at a no-load pace, since the speed will invariably decrease as the load amount increases. The rate the speed will decrease will directly correlate with the amount of force and the initial speed.

Operating Conditions

Actuators are commonly rated using the standard IP Code rating system. Those that are rated for dangerous environments will have a higher IP rating than those for personal or common industrial use.

Durability

This will be determined by each individual manufacturer, depending on usage and quality.

References

- Bănică, Florinel-Gabriel (2012). Chemical Sensors and Biosensors:Fundamentals and Applications. Chichester, UK: John Wiley & Sons. p. 576. ISBN 978-1-118-35423-0.

- "What's the Difference Between Pneumatic, Hydraulic, and Electrical Actuators?". machinedesign.com. Retrieved 2016-04-26.

System Architecture: An Overview

System architecture deals with the design of systems that store content or data. It may use different computer languages or architecture description languages to access its content. This chapter gives an in-depth understanding of system architecture, and provides the reader with an elucidated knowledge on the subject matter.

Systems Architecture

A system architecture or systems architecture is the conceptual model that defines the structure, behavior, and more views of a system. An architecture description is a formal description and representation of a system, organized in a way that supports reasoning about the structures and behaviors of the system.

A system architecture can comprise system components, the expand systems developed, that will work together to implement the overall system. There have been efforts to formalize languages to describe system architecture, collectively these are called architecture description languages (ADLs).

Overview

Various organizations can define systems architecture in different ways, including:

- The fundamental organization of a system, embodied in its components, their relationships to each other and to the environment, and the principles governing its design and evolution.

- A representation of a system, including a mapping of functionality onto hardware and software components, a mapping of the software architecture onto the hardware architecture, and human interaction with these components.

- An allocated arrangement of physical elements which provides the design solution for a consumer product or life-cycle process intended to satisfy the requirements of the functional architecture and the requirements baseline.

- An architecture comprises the most important, pervasive, top-level, strategiinventions, decisions, and their associated rationales about the overall structure (i.e., essential elements and their relationships) and associated characteristics and behavior.

- A description of the design and contents of a computer system. If documented, it may include information such as a detailed inventory of current hardware, software and networking capabilities; a description of long-range plans and priorities for future purchases, and a plan for upgrading and/or replacing dated equipment and software.

- A formal description of a system, or a detailed plan of the system at component level to guide its implementation.

- The composite of the design architectures for products and their life-cycle processes.

- The structure of components, their interrelationships, and the principles and guidelines governing their design and evolution over time.

One can think of system architecture as a set of representations of an existing (or future) system. These representations initially describe a general, high-level functional organization, and are progressively refined to more detailed and concrete descriptions.

System architecture conveys the informational content of the elements comprising a system, the relationships among those elements, and the rules governing those relationships. The architectural components and set of relationships between these components that an architecture description may consist of hardware, software, documentation, facilities, manual procedures, or roles played by organizations or people.

A system architecture primarily concentrates on the internal interfaces among the system's components or subsystems, and on the interface(s) between the system and its external environment, especially the user. (In the specificase of computer systems, this latter, special, interface is known as the computer human interface, *AKA* human computer interface, or CHI; formerly called the man-machine interface.)

One can contrast a system architecture with system architecture engineering (SAE) - the method and discipline for effectively implementing the architecture of a system:

- SAE is a *method* because a sequence of steps is prescribed to produce or to change the architecture of a system within a set of constraints.

- SAE is a *discipline* because a body of knowledge is used to inform practitioners as to the most effective way to architect the system within a set of constraints.

History

Systems architecture depends heavily on practices and techniques which were developed over thousands of years in many other fields, perhaps the most important being civil architecture.

- Prior to the advent of digital computers, the electronics and other engineering disciplines used the term "system" as it is still commonly used today. However, with the arrival of digital computers and the development of software engineering as a separate discipline, it was often necessary to distinguish among engineered hardware artifacts, software artifacts, and the combined artifacts. A programmable hardware artifact, or computing machine, that lacks its computer program is impotent; even as a software artifact, or program, is equally impotent unless it can be used to alter the sequential states of a suitable (hardware) machine. However, a hardware machine and its programming can be designed to perform an almost illimitable number of abstract and physical tasks. Within the computer and software engineering disciplines (and, often, other engineering disciplines, such as communications), then, the term system came to be defined as containing all of the elements necessary (which generally includes both hardware and software) to perform a useful function.

- Consequently, within these engineering disciplines, a system generally refers to a programmable hardware machine and its included program. And a systems engineer is defined as one concerned with the complete device, both hardware and software and, more particularly, all of the interfaces of the device, including that between hardware and software, and especially between the complete device and its user (the CHI). The hardware engineer deals (more or less) exclusively with the hardware device; the software engineer deals (more or less) exclusively with the computer program; and the systems engineer is responsible for seeing that the program is capable of properly running within the hardware device, and that the system composed of the two entities is capable of properly interacting with its external environment, especially the user, and performing its intended function.

- A systems architecture makes use of elements of both software and hardware and is used to enable design of such a composite system. A good architecture may be viewed as a 'partitioning scheme,' or algorithm, which partitions all of the system's present and foreseeable requirements into a workable set of cleanly bounded subsystems with nothing left over. That is, it is a partitioning scheme which is exclusive, inclusive, and exhaustive. A major purpose of the partitioning is to arrange the elements in the sub systems so that there is a minimum of interdependencies needed among them. In both software and hardware, a good sub system tends to be seen to be a meaningful "object". Moreover, a good architecture provides for an easy mapping to the user's requirements and the validation tests of the user's requirements. Ideally, a mapping also exists from every least element to every requirement and test.

Types

Several types of systems architectures (underlain by the same fundamental principles) have been identified as follows:

- Hardware architecture

- Software architecture

- Enterprise architecture

- Collaborative systems architectures(such as the Internet, intelligent transportation systems, and joint air defense systems)

- Manufacturing systems architectures

- Strategisystems architecture

Types of System Architectures

Hardware Architecture

In engineering, hardware architecture refers to the identification of a system's physical components and their interrelationships. This description, often called a hardware design model, allows hardware designers to understand how their components fit into a system architecture and provides to software component designers important information needed for software development and integration. Clear definition of a hardware architecture allows the various traditional engineering disciplines (e.g., electrical and mechanical engineering) to work more effectively together to develop and manufacture new machines, devices and components.

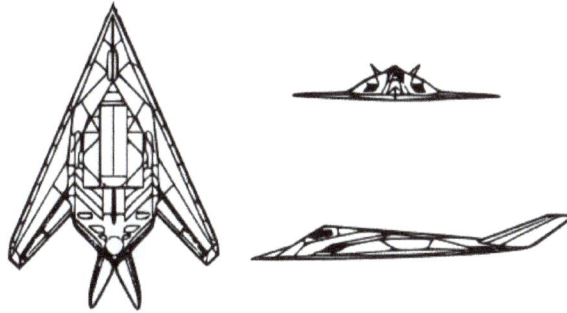

An orthographically projected diagram of the F-117A Nighthawk.

An F-117 conducts a live exercise bombing run using GBU-27 laser-guided bombs.

Hardware is also an expression used within the computer engineering industry to explicitly distinguish the (electronicomputer) hardware from the *software* that runs on it. But *hardware,* within the automation and software engineering disciplines, need not simply be a computer of some sort. A modern automobile runs vastly more *software* than the Apollo spacecraft. Also, modern aircraft cannot function without running tens of millions of computer instructions embedded and distributed throughout the aircraft and resident in both standard computer hardware and in specialized hardward components such as IC wired logigates, analog and hybrid devices, and other digital components. The need to effectively model how separate physical components combine to form complex systems is important over a wide range of applications, including computers, personal digital assistants (PDAs), cell phones, surgical instrumentation, satellites, and submarines.

Hardware architecture is the representation of an engineered (or *to be engineered*) electronior electromechanical hardware system, and the process and discipline for effectively implementing the design(s) for such a system. It is generally part of a larger integrated system encompassing information, software, and device prototyping.

It is a *representation* because it is used to convey information about the related elements comprising a hardware system, the relationships among those elements, and the rules governing those relationships.

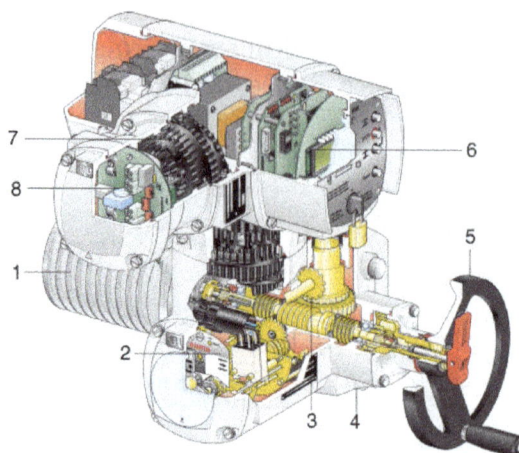

Electrimulti-turn actuator with controls.

It is a *process* because a sequence of steps is prescribed to produce or change the architecture, and/or a design from that architecture, of a hardware system within a set of constraints.

It is a *discipline* because a body of knowledge is used to inform practitioners as to the most effective way to design the system within a set of constraints.

A hardware architecture is primarily concerned with the internal electrical (and, more rarely, the mechanical) interfaces among the system's components or subsystems, and the interface between the system and its external environment, especially the devices operated by or the electronidisplays viewed by a user. (This latter, special interface, is known as the computer human interface, *AKA* human computer interface, or HCI; formerly called the man-machine interface.) Integrated circuit (IC) designers are driving current technologies into innovative approaches for new products. Hence, multiple layers of active devices are being proposed as single chip, opening up opportunities for disruptive microelectronic, optoelectronic, and new microelectromechanical hardware implementation.

Background

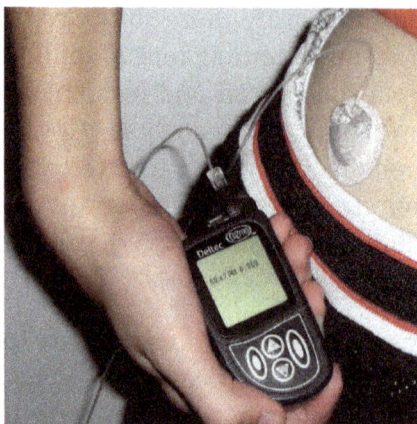

A hardware architecture example, which is integrated as a handheld medical device for diabetes monitoring.

Prior to the advent of digital computers, the electronics and other engineering disciplines used the terms system and hardware as they are still commonly used today. However, with the arrival of digital computers on the scene and the development of software engineering as a separate discipline, it was often necessary to distinguish among engineered *hardware* artifacts, *software* artifacts, and the combined artifacts.

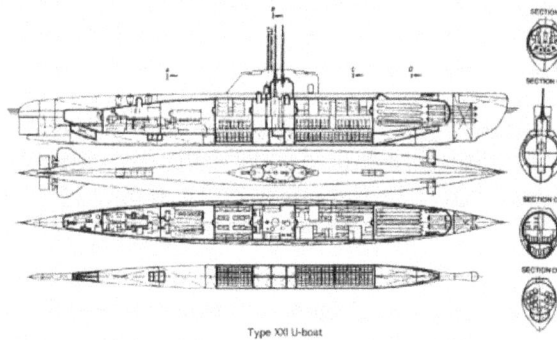

U-Boat layout, with detailed equipment hardware specification and functionality.

A programmable hardware artifact, or machine, that lacks its computer program is impotent; even as a software artifact, or program, is equally impotent unless it can be used to alter the sequential states of a suitable (hardware) machine. However, a hardware machine and its programming can be designed to perform an almost illimitable number of abstract and physical tasks. Within the computer and software engineering disciplines (and, often, other engineering disciplines, such as communications), then, the terms hardware, software, and system came to distinguish between the hardware that runs a computer program, the software, and the hardware device complete with its program.

The *hardware* engineer or architect deals (more or less) exclusively with the hardware device; the *software* engineer or architect deals (more or less) exclusively with the program; and the *systems* engineer or systems architect is responsible for seeing that the programming is capable of properly running within the hardware device, and that the system composed of the two entities is capable of properly interacting with its external environment, especially the user, and performing its intended function.

A hardware architecture, then, is an abstract representation of an electronior an electromechanical device capable of running a fixed or changeable program.

A hardware architecture generally includes some form of analog, digital, or hybrid electronicomputer, along with electroniand mechanical sensors and actuators. Hardware design may be viewed as a 'partitioning scheme,' or algorithm, which considers all of the system's present and foreseeable requirements and arranges the necessary hardware components into a workable set of cleanly bounded subsystems with no more parts than are required. That is, it is a partitioning scheme that is exclusive, inclusive, and exhaustive. A major purpose of the partitioning is to arrange the elements in the hardware subsystems so that there is a minimum of electrical connections and electronicommunications needed among them. In both software and hardware, a good subsystem tends to be seen as a meaningful "object." Moreover, a clear allocation of user requirements to the architecture (hardware and software) provides an effective basis for validation tests of the user's requirements in the as-built system.

Software architecture

Software architecture refers to the fundamental structures of a software system, the discipline of creating such structures, and the documentation of these structures. These structures are needed to reason about the software system. Each structure comprises software elements, relations among them, and properties of both elements and relations, along with rationale for the introduction and configuration of each element. The *architecture* of a software system is a metaphor, analogous to the architecture of a building.

Software architecture is about making fundamental structural choices which are costly to change once implemented. Software architecture choices, also called architectural decisions, include specific structural options from possibilities in the design of software. For example, the systems that controlled the space shuttle launch vehicle had the requirement of being very fast and very reliable. Therefore, an appropriate real-time computing language would need to be chosen. Additionally, to satisfy the need for reliability the choice could be made to have multiple redundant and independently produced copies of the program, and to run these copies on independent hardware while cross-checking results.

Documenting software architecture facilitates communication between stakeholders, captures decisions about the architecture design, and allows reuse of design components between projects.

Scope

Opinions vary as to the scope of software architectures:

- *Overall, macroscopic system structure*; this refers to architecture as a higher level abstraction of a software system that consists of a collection of computational *components* together with *connectors* that describe the interaction between these components.

- *The important stuff—whatever that is*; this refers to the fact that software architects should concern themselves with those decisions that have high impact on the system and its stakeholders.

- *That which is fundamental to understanding a system in its environment*; in this definition, the environment is characterized by *stakeholder concerns*, technical constraints, and various dimensions of project context.

- *Things that people perceive as hard to change*; since designing the architecture often takes place at the beginning of a software system's lifecycle, the architect should focus on decisions that "have to" be right the first time. Following this line of thought, architectural design issues may become non-architectural once their irreversibility can be overcome.

- *A set of architectural design decisions*; software architecture should not be considered merely a set of models or structures, but should include the decisions that lead to these particular structures, and the rationale behind them (e.g., justifications, answers to "why" questions)). This insight has led to substantial research into software architecture knowledge management.

There is no sharp distinction between software architecture versus design and requirements engineering. They are all part of a "chain of intentionality" from high-level intentions to low-level details. This duality is also referred to as the "twin peaks" of software engineering.

Characteristics

Software architecture exhibits the following:

Multitude of stakeholders: software systems have to cater to a variety of stakeholders such as business managers, application owners, developers, end users and infrastructure operators. These stakeholders all have their own concerns with respect to the system. Balancing these concerns and demonstrating how they are addressed is part of designing the system. This implies that architecture involves dealing with a broad variety of concerns and stakeholders, and has a multidisciplinary nature. Software architect require non-technicals skills such as communication and negotiation competencies.

Separation of concerns: the established way for architects to reduce complexity is to separate the concerns that drive the design. Architecture documentation shows that all stakeholder concerns are addressed by modeling and describing the architecture from separate points of view associated with the various stakeholder concerns. These separate descriptions are called architectural views (see for example the 4+1 Architectural View Model).

Quality-driven: classisoftware design approaches (e.g. Jackson Structured Programming) were driven by required functionality and the flow of data through the system, but the current insightis that the architecture of a software system is more closely related to its quality attributes such as fault-tolerance, backward compatibility, extensibility, reliability, maintainability, availability, security, usability, and other such –ilities. Stakeholder concerns often translate into requirements and constraints on these quality attributes, which are variously called non-functional requirements, extra-functional requirements, behavioral requirements, or quality attribute requirements.

Recurring styles: like building architecture, the software architecture discipline has developed standard ways to address recurring concerns. These "standard ways" are called by various names at various levels of abstraction. Common terms for recurring solutions are architectural style, principle,reference architecture and architectural pattern.

Conceptual integrity: a term introduced by Fred Brooks in The Mythical Man-Month to denote the idea that the architecture of a software system represents an overall vision of what it should do and how it should do it. This vision should be separated from its implementation. The architect assumes the role of "keeper of the vision", making sure that additions to the system are in line with the architecture, hence preserving conceptual integrity.

Motivation

Software architecture is an "intellectually graspable" abstraction of a complex system. This abstraction provides a number of benefits:

- *It gives a basis for analysis of software systems' behavior before the system has been built.* The ability to verify that a future software system fulfills its stakeholders' needs without actually having to build it represents substantial cost-saving and risk-mitigation. A number of techniques have been developed in academia and practice to perform such analyses, for instance ATAM, ARID and TARA.

- *It provides a basis for re-use of elements and decisions.* A complete software architecture or parts of it, like individual architectural strategies and decisions, can be re-used across multiple systems whose stakeholders require similar quality attributes or functionality, saving design costs and mitigating the risk of design mistakes (assuming that the project contexts match).

- *It supports early design decisions that impact a system's development, deployment, and maintenance life.* Getting the early, high-impact decisions right is important to prevent schedule and budget overruns. On the other hand, a principle of lean software development is to defer decisions until the last responsible moment (M. and T. Poppendieck); however, it is not always clear when the this moment for a particular subset of decisions has come.

- *It facilitates communication with stakeholders, contributing to a system that better fulfills their needs.* Communicating about complex systems from the point of view of stakeholders helps them understand the consequences of their stated requirements and the design decisions based on them. Architecture gives the ability to communicate about design decisions before the system is implemented, when they are still relatively easy to adapt.

- *It helps in risk management.* Software architecture helps to reduce risks and chance of failure.

- *It enables cost reduction.* Software architecture is a means to manage risk and costs in complex IT projects.

History

The comparison between software design and (civil) architecture was first drawn in the late 1960s , but the term *software architecture* became prevalent only in the 1990s. The field of computer science had encountered problems associated with complexity since its formation. Earlier problems of complexity were solved by developers by choosing the right data structures, developing algorithms, and by applying the concept of separation of concerns. Although the term "software architecture" is relatively new to the industry, the fundamental principles of the field have been applied sporadically by software engineering pioneers since the mid-1980s. Early attempts to capture and explain software architecture of a system were imprecise and disorganized, often characterized by a set of box-and-line diagrams.

Software architecture as a concept has its origins in the research of Edsger Dijkstra in 1968 and David Parnas in the early 1970s. These scientists emphasized that the structure of a software system matters and getting the structure right is critical. During the 1990s there was a concerted effort to define and codify fundamental aspects of the discipline, with research work concentrating on architectural styles (patterns), architecture description languages, and architecture documentation. Research institutions have played a prominent role in furthering software architecture as a discipline. For instance, Mary Shaw and David Garlan of Carnegie Mellon wrote a book titled *Soft-*

ware Architecture: Perspectives on an Emerging Discipline in 1996, which promoted software architecture concepts such as components, connectors, and styles.

IEEE 1471-2000, *Recommended Practice for Architecture Description of Software-Intensive Systems*, was the first formal standard in the area of software architecture. It was adopted in 2007 by ISO as ISO/IEC 42010:2007. In November 2011, IEEE 1471–2000 was superseded by ISO/IEC/IEEE 42010:2011, *Systems and software engineering — Architecture description* (jointly published by IEEE and ISO). While in IEEE 1471, software architecture was about the architecture of "software-intensive systems", defined as "any system where software contributes essential influences to the design, construction, deployment, and evolution of the system as a whole", the 2011 edition goes a step further by including the ISO/IEC 15288 and ISO/IEC 12207 definitions of a system, which embrace not only hardware and software, but also "humans, processes, procedures, facilities, materials and naturally occurring entities".

Architecture Activities

There are many activities that a software architect performs. A software architect typically works with project managers, discusses architecturally significant requirements with stakeholders, designs a software architecture, evaluates a design, communicates with designers and stakeholders, documents the architectural design and more. There are four core activities in software architecture design. These core architecture activities are performed iteratively and at different stages of the initial software development life-cycle, as well as over the evolution of a system.

Architectural Analysis is the process of understanding the environment in which a proposed system or systems will operate and determining the requirements for the system. The input or requirements to the analysis activity can come from any number of stakeholders and include items such as:

- what the system will do when operational (the functional requirements)

- how well the system will perform runtime non-functional requirements such as reliability, operability, performance efficiency, security, compatibility defined in ISO/IEC 25010:2011 standard

- development-time non-functional requirements such as maintainability and transferability defined in ISO 25010:2011 standard

- business requirements and environmental contexts of a system that may change over time, such as legal, social, financial, competitive, and technology concerns

The outputs of the analysis activity are those requirements that have a measurable impact on a software system's architecture, called architecturally significant requirements.

Architectural Synthesis or design is the process of creating an architecture. Given the architecturally significant requirements determined by the analysis, the current state of the design and the results of any evaluation activities, the design is created and improved.

Architecture Evaluation is the process of determining how well the current design or a portion of it satisfies the requirements derived during analysis. An evaluation can occur whenever an architect is considering a design decision, it can occur after some portion of the design has been completed, it can occur after the final design has been completed or it can occur after the system has been constructed. Some of the available software architecture evaluation techniques include Architecture Tradeoff Analysis Method (ATAM) and TARA. Frameworks for comparing the techniques are discussed in and.

Architecture Evolution is the process of maintaining and adapting an existing software architecture to meet requirement and environmental changes. As software architecture provides a fundamental structure of a software system, its evolution and maintenance would necessarily impact its fundamental structure. As such, architecture evolution is concerned with adding new functionality as well as maintaining existing functionality and system behaviour, for instance, via architectural refactoring.

Architecture requires critical supporting activities. These supporting activities take place throughout the core software architecture process. They include knowledge management and communication, design reasoning and decision making, and documentation.

Architecture Supporting Activities

Software architecture supporting activities are carried out during core software architecture activities. These supporting activities assist a software architect to carry out analysis, synthesis, evaluation and evolution. For instance, an architect has to gather knowledge, make decisions and document during the analysis phase.

- Knowledge Management and Communication is the activity of exploring and managing knowledge that is essential to designing a software architecture. A software architect does not work in isolation. They get inputs, functional and non-functional requirements and design contexts, from various stakeholders; and provides outputs to stakeholders. Software architecture knowledge is often tacit and is retained in the heads of stakeholders. Software Architecture Knowledge Management (AKM) is about finding, communicating, and retaining knowledge. As software architecture design issues are intricate and interdependent, a knowledge gap in design reasoning can lead to incorrect software architecture design. Examples of AKM and communication activities include searching for design patterns, prototyping, asking experienced developers and architects, evaluating the designs of similar systems, sharing knowledge with other designers and stakeholders.

- Design Reasoning and Decision Making is the activity of evaluating design decisions. This activity is fundamental to all three core software architecture activities. It entails gathering and associating decision contexts, formulating design decision problems, finding solution options and evaluating tradeoffs before making decisions. This process occurs at different levels of decision granularity, while evaluating significant architectural requirements and software architecture decisions, and software architecture analysis, synthesis, and evaluation. Examples of reasoning activities include understanding the impacts of a requirement or a design on quality attributes, questioning the issues that a design might cause, assessing possible solution options, and evaluating the tradeoffs between solutions.

- Documentation is the activity of recording the design generated during the software architecture process. A system design is described using several views that frequently include a statiview showing the code structure of the system, a dynamiview showing the actions of the system during execution, and a deployment view showing how a system is placed on hardware for execution. Kruchten's 4+1 view suggests a description of commonly used views for documenting software architecture; Documenting Software Architectures: Views and Beyond has descriptions of the kinds of notations that could be used within the view description. Examples of documentation activities are writing a specification, recording a system design model, documenting a design rationale, developing a viewpoint, documenting views. Software engineering methods such as the OpenUP and architecture design methods such as The Process of Software Architecting (P. Eeles, P. Cripps) suggest artifact (a.k.a. work product) types and templates for these documentation activities; ISO/IEC/IEEE 42010:2011 is accompanied by a documentation template as well (http://www.iso-architecture.org/ieee-1471/templates/).

Software Architecture Topics

Software Architecture Description

Software architecture description involves the principles and practices of modeling and representing architectures, using mechanisms such as: architecture description languages, architecture viewpoints, and architecture frameworks.

Architecture Description Languages

An architecture description language (ADL) is any means of expression used to describe a software architecture (ISO/IEC/IEEE 42010). Many special-purpose ADLs have been developed since the 1990s, including ArchiMate, AADL (SAE standard), Wright (developed by Carnegie Mellon), Acme (developed by Carnegie Mellon), xADL (developed by UCI), Darwin (developed by Imperial College London), DAOP-ADL (developed by University of Málaga), SBC-ADL (developed by National Sun Yat-Sen University), and ByADL (University of L'Aquila, Italy).

ADLs have not yet succeeded on a broad scale in practice; UML, often profiled, and Informal Rich Pictures (IPRs) a.k.a. box-and-line diagrams dominate. Usage of UML has been criticized by some thought leaders, but successes have also been reported. Simon Brown's Context, Containers, Components, Classes (C4) model is a recent adiditon to the architect's notation toolbox: https://www.voxxed.com/blog/2014/10/simple-sketches-for-diagramming-your-software-architecture/.

According to Gregor Hohpe, architects should stop drawing diagrams, in whatever notation, and start communicating: http://www.enterpriseintegrationpatterns.com/ramblings/94_37things.html.

Architecture Viewpoints

Software architecture descriptions are commonly organized into views, which are analogous to the different types of blueprints made in building architecture. Each view addresses a set of system concerns, following the conventions of its *viewpoint*, where a viewpoint is a specification that describes the notations, modeling and analysis techniques to use in a view that express the architecture in question from the perspective of a given set of stakeholders and their concerns (ISO/IEC/

IEEE 42010). The viewpoint specifies not only the concerns framed (i.e., to be addressed) but the presentation, model kinds used, conventions used and any consistency (correspondence) rules to keep a view consistent with other views.

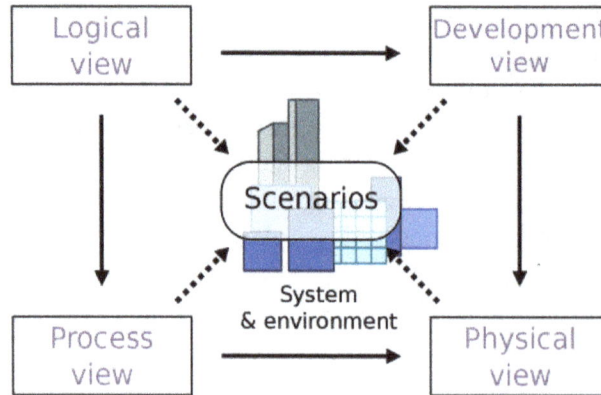

4+1 Architectural View Model.

Popular viewpoint models include the 4+1 views on software architecture, the viewpoints and per-spectices catalog by Nick Rozanski and Eoin Woods, and the IBM ADS viewpoint model by Phlippe Spaas et al.

Architecture Frameworks

An architecture framework captures the "conventions, principles and practices for the description of architectures established within a specifidomain of application and/or community of stake-holders" (ISO/IEC/IEEE 42010). A framework is usually implemented in terms of one or more viewpoints or ADLs.

Architecture Design Methods

Methods define process models (activities performed by roles) and specify artifacts to be created and delivered; they may also suggest technqiues and practices that assist practitioners when per-forming the activities and producing the artifacts defined in the method. Five such methods are compared and consolidated in.

Software Architecture and Agile Development

The importance of architecture was stated in the early works on Agile; for instance, Ken Schwaber's original Scrum paper from OOPSLA '97 has the notion of a pregame, in which a high-level system architecture is established (http://www.jeffsutherland.org/oopsla/schwapub. pdf). However, there are also concerns that software architecture leads to too much Big Design Up Front, especially among proponents of Agile software development. A number of methods have been developed to balance the trade-offs of up-front design and agility, including the agile method DSDM which mandates a "Foundations" phase during which "just enough" architectural foundations are laid. IEEE Software devoted a special issue to the interaction between agility and architecture. P. Kruchten, one of the creators of the Unified Process (UP) and the original

4+1 views on software architecture, summarizes the synergetirelationship in a December 2013 blog post called "Agile architecture".

Architectural Styles and Patterns

An architectural pattern is a general, reusable solution to a commonly occurring problem in software architecture within a given context. Architectural patterns are often documented as software design patterns.

Following traditional building architecture, a 'software architectural style' is a specifimethod of construction, characterized by the features that make it notable" (Architectural style). "An architectural style defines: a family of systems in terms of a pattern of structural organization; a vocabulary of components and connectors, with constraints on how they can be combined." "Architectural styles are reusable 'packages' of design decisions and constraints that are applied to an architecture to induce chosen desirable qualities."

There are many recognized architectural patterns and styles, among them:

- Blackboard

- Client-server (2-tier, 3-tier, n-tier, cloud computing exhibit this style)

- Component-based

- Data-centric

- Event-driven (or Implicit invocation)

- Layered (or Multilayered architecture)

- Monolithiapplication

- Peer-to-peer (P2P)

- Pipes and filters

- Plug-ins

- Representational state transfer (REST)

- Rule-based

- Service-oriented architecture and microservices as its implementation approach

- Shared nothing architecture

- Space-based architecture

Some software architects treat architectural patterns and architectural styles as the same, Others treat styles as compositions of patterns combined with architectural principles that jointly address a particular *intent*. These two positions have in common that both patterns and styles are idioms for architects to use; they "provide a common language" or "vocabulary" with which to describe classes of systems.

Software Architecture Erosion

Software architecture erosion (or "decay") refers to the gap observed between the planned and actual architecture of a software system as realized in its implementation. Software architecture erosion occurs when implementation decisions either do not fully achieve the architecture-as-planned or otherwise violate constraints or principles of that architecture. The gap between planned and actual architectures is sometimes understood in terms of the notion of technical debt.

As an example, consider a strictly layered system, where each layer can only use services provided by the layer immediately below it. Any source code component that does not observe this constraint represents an architecture violation. If not corrected, such violations can transform the architecture into a monolithiblock, with adverse effects on understandability, maintainability, and evolvability.

Various approaches have been proposed to address erosion. "These approaches, which include tools, techniques and processes, are primarily classified into three genericategories that attempt to minimise, prevent and repair architecture erosion. Within these broad categories, each approach is further broken down reflecting the high-level strategies adopted to tackle erosion. These are: process-oriented architecture conformance, architecture evolution management, architecture design enforcement, architecture to implementation linkage, self-adaptation and architecture restoration techniques consisting of recovery, discovery and reconciliation."

There are two major techniques to detect architectural violations: reflexion models and domain-specifilanguages. Reflexion model (RM) techniques compare a high-level model provided by the system's architects with the source code implementation. Examples of commercial RM-based tools include the Bauhaus Suite (developed by Axivion), SAVE (developed by Fraunhofer IESE) and Structure-101 (developed by Headway Software). There are also domain-specifilanguages with focus on specifying and checking architectural constraints, including .QL (developed by Semmle Limited) and DCL (from Federal University of Minas Gerais).

Software Architecture Recovery

Software architecture recovery (or reconstruction, or reverse engineering) includes the methods, techniques and processes to uncover a software system's architecture from available information, including its implementation and documentation. Architecture recovery is often necessary to make informed decisions in the face of obsolete or out-of-date documentation and architecture erosion: implementation and maintenance decisions diverging from the envisioned architecture.

Design

Architecture is design but not all design is architectural. In practice, the architect is the one who draws the line between software architecture (architectural design) and detailed design (non-architectural design). There are no rules or guidelines that fit all cases, although there have been attempts to formalize the distinction. According to the *Intension/Locality Hypothesis*, the distinction between architectural and detailed design is defined by the *Locality Criterion*, according to which a statement about software design is non-local (architectural) if and only if a program that satisfies it can be expanded into a program that does not. For example, the client–server style is architectural (strategic) because a program that is built on this principle can be expanded into a program that is not client–server—for example, by adding peer-to-peer nodes.

Requirements Engineering

Requirements engineering and software architecture can be seen as complementary approaches: while software architecture targets the 'solution space' or the 'how', requirements engineering addresses the 'problem space' or the 'what'. Requirements engineering entails the elicitation, negotiation, specification, validation, documentation and management of requirements. Both requirements engineering and software architecture revolve around stakeholder concerns, needs and wishes.

There is considerable overlap between requirements engineering and software architecture, as evidenced for example by a study into five industrial software architecture methods that concludes that *"the inputs (goals, constrains, etc.) are usually ill-defined, and only get discovered or better understood as the architecture starts to emerge"* and that while *"most architectural concerns are expressed as requirements on the system, they can also include mandated design decisions"*. In short, the choice of required behavior given a particular problem impacts the architecture of the solution that addresses that problem, while at the same time the architectural design may impact the problem and introduce new requirements. Approaches such as the Twin Peaks model aim to exploit the synergistirelation between requirements and architecture.

Other Types of 'Architecture'

Computer architecture

> Computer architecture targets the internal structure of a computer system, in terms of collaborating hardware components such as the CPU – or processor – the bus and the memory.

Systems architecture

> The term systems architecture has originally been applied to the architecture of systems that consists of both hardware and software. The main concern addressed by the systems architecture, also known as IT architecture, is then the integration of software and hardware in a complete, correctly working device. In another common – much broader – meaning, the term applies to the architecture of any complex system which may be of technical, sociotechnical or social nature.

Enterprise architecture

> The goal of enterprise architecture is to "translate business vision and strategy into effective enterprise". Enterprise architecture frameworks, such as TOGAF and the Zachman Framework, usually distinguish between different enterprise architecture layers. Although terminology differs from framework to framework, many include at least a distinction between a *business layer*, an *application* (or *information*) *layer*, and a *technology layer*. Enterprise architecture addresses among others the alignment between these layers, usually in a top-down approach. Continuing the building architecture metaphor for software architecture, enterprise architecture is analogous to city-level planning.

Enterprise Architecture

Enterprise architecture (EA) is "a well-defined practice for conducting enterprise analysis, design, planning, and implementation, using a holistiapproach at all times, for the successful development and execution of strategy. Enterprise architecture applies architecture principles and practices to guide organizations through the business, information, process, and technology changes necessary to execute their strategies. These practices utilize the various aspects of an enterprise to identify, motivate, and achieve these changes."

Practitioners of enterprise architecture, *enterprise architects*, are responsible for performing the analysis of business structure and processes and are often called upon to draw conclusions from the information collected to address the goals of enterprise architecture: effectiveness, efficiency, agility, and durability.

Overview

In the enterprise architecture literature and community, there are various perspectives in regards to the meaning of the term *enterprise architecture*. As of 2012, no official definition exists; rather, various organizations (publiand private) promote their understanding of the term. Consequently, the enterprise architecture literature offers many definitions for the term enterprise architecture; some of which are complementary, others nuances, and others yet are in opposition.

The MIT Center for Information Systems Research (MIT CISR) in 2007 defined enterprise architecture as the specifiaspects of a business that are under examination:

> *Enterprise architecture is the organizing logifor business processes and IT infrastructure reflecting the integration and standardization requirements of the company's operating model. The operating model is the desired state of business process integration and business process standardization for delivering goods and services to customers.*

The Enterprise Architecture Body of Knowledge defines enterprise architecture as a practice, which

> *analyzes areas of common activity within or between organizations, where information and other resources are exchanged to guide future states from an integrated viewpoint of strategy, business and technology.*

IT analysis firm Gartner defines the term as a discipline where an enterprise is led through change. According to their glossary,

> "Enterprise architecture (EA) is a discipline for proactively and holistically leading enterprise responses to disruptive forces by identifying and analyzing the execution of change toward desired business vision and outcomes. EA delivers value by presenting business and IT leaders with signature-ready recommendations for adjusting policies and projects to achieve target business outcomes that capitalize on relevant business disruptions. EA is used to steer decision making toward the evolution of the future state architecture."

Each of the definitions above underplay the historical reality that enterprise architecture emerged from methods for documenting and planning information systems architectures, and the current reality that most enterprise architecture practitioners report to a CIO or other IT department manager. In a business organization structure today, the enterprise architecture team performs an ongoing business function that helps business and IT managers to figure out the best strategies to support and enable business development and business change – in relation to the business information systems the business depends on.

Enterprise Architecture Topics

The Terms Enterprise and Architecture

The term *enterprise* can be defined as describing an organizational unit, organization, or collection of organizations that share a set of common goals and collaborate to provide specifiproducts or services to customers.

In that sense, the term enterprise covers various types of organizations, regardless of their size, ownership model, operational model, or geographical distribution. It includes those organizations' complete socio-technical systems, including people, information, processes and technologies.

The term *architecture* refers to fundamental concepts or properties of a system in its environment embodied in its elements, relationships, and in the principles of its design and evolution.

An enterprise, understood as a socio-technical system, defines the scope of the enterprise architecture.

Scope of Enterprise Architecture

Current perspectives, or beliefs, held by enterprise architecture practitioners and scholars, with regards to the meaning of the enterprise architecture, typically gravitate towards one or a hybrid of three schools of thought:

1. Enterprise IT design – the purpose of EA is the greater alignment between IT and business concerns. The main purpose of enterprise architecture is to guide the process of planning and design the IT/IS capabilities of an enterprise in order to meet desired organizational objectives. Typically, architecture proposals and decisions are limited to the IT/IS aspects of the enterprise; other aspects only serve as inputs.

2. Enterprise integrating – According to this school, the purpose of EA is to achieve greater coherency between the various concerns of an enterprise (HR, IT, Operations, etc.) including the linking between strategy formulation and execution. Typically, architecture proposals and decisions encompass all the aspects of the enterprise.

3. Enterprise ecological adaptation – the purpose of EA is to foster and maintain the learning capabilities of enterprises so that they may be sustainable. Consequently, a great deal of emphasis is put on improving the capabilities of the enterprise to improve itself, to innovate and to coevolve with its environment. Typically, proposals and decisions encompass both the enterprise and its environment.

One's belief with regards to the meaning of enterprise architecture will impact how one sees its purpose, its scope, the means of achieving it, the skills needed to conduct it, and the locus of responsibility for conducting it

Architectural Description of an Enterprise

According to the standard ISO/IEC/IEEE 42010, the product used to describe the architecture of a system is called an *architectural description*. In practice, an architectural description contains a variety of lists, tables and diagrams. These are models known as *views*. In the case of Enterprise Architecture, these models describe the logical business functions or capabilities, business processes, human roles and actors, the physical organization structure, data flows and data stores, business applications and platform applications, hardware and communications infrastructure.

The UK National Computing Centre EA best practice guidance states

Normally an EA takes the form of a comprehensive set of cohesive models that describe the structure and functions of an enterprise... The individual models in an EA are arranged in a logical manner that provides an ever-increasing level of detail about the enterprise.

The architecture of an enterprise is described with a view to improving the manageability, effectiveness, efficiency or agility of the business, and ensuring that money spent on information technology (IT) is justified.

Paramount to *changing* the enterprise architecture is the identification of a sponsor, his/her mission, vision and strategy and the governance framework to define all roles, responsibilities and relationships involved in the anticipated transformation. Changes considered by enterprise architects typically include:

- innovations in the structure or processes of an organization

- innovations in the use of information systems or technologies

- the integration and/or standardization of business processes,

- improving the quality and timeliness of business information.

A methodology for developing and using architecture to guide the transformation of a business from a baseline state to a target state, sometimes through several transition states, is usually known as an enterprise architecture framework. A framework provides a structured collection of processes, techniques, artifact descriptions, reference models and guidance for the production and use of an enterprise-specifiarchitecture description.

Benefits of Enterprise Architecture

The benefits of enterprise architecture are achieved through its direct and indirect contributions to organizational goals. It has been found that the most notable benefits of enterprise architecture can be observed in the following areas:

- Organizational design - Enterprise architecture provides support in the areas related to design and re-design of the organizational structures during mergers, acquisitions or during general organizational change.

- Organizational processes and process standards - Enterprise architecture helps enforce discipline and standardization of business processes, and enable process consolidation, reuse and integration.

- Project portfolio management - Enterprise architecture supports investment decision-making and work prioritization.

- Project management - Enterprise architecture enhances the collaboration and communication between project stakeholders. Enterprise architecture contributes to efficient project scoping, and to definition of more complete and consistent project deliverables.

- Requirements Engineering - Enterprise architecture increases the speed of requirement elicitation and the accuracy of requirement definitions, through publishing of the enterprise architecture documentation.

- System development - Enterprise architecture contributes to optimal system designs and efficient resource allocation during system development and testing.

- IT management and decision making - Enterprise architecture is found to help enforce discipline and standardization of IT planning activities and to contribute to reduction in time for technology-related decision making.

- IT value - Enterprise architecture helps reduce the systems implementation and operational costs, and minimize replication of IT infrastructure services across business units.

- IT complexity - Enterprise architecture contributes to reduction in IT complexity, consolidation of data and applications, and to better interoperability of the systems.

- IT openness - Enterprise architecture contributes to more open and responsive IT as reflected through increased accessibility of data for regulatory compliance, and increased transparency of infrastructure changes.

- IT risk management - Enterprise architecture contributes to reduction of business risk from system failures and security breaches. Enterprise architecture helps reduce risks of project delivery.

Examples of Enterprise Architecture use

Documenting the architecture of enterprises is done within the U.S. Federal Government in the context of the Capital Planning and Investment Control (CPIC) process.

The Federal Enterprise Architecture (FEA) reference models guides federal agencies in the development of their architectures.

Companies such as Independence Blue Cross, Intel, Volkswagen AG and InterContinental Hotels Group use enterprise architecture to improve their business architectures as well as to improve business performance and productivity.

For various understandable reasons, commercial organizations rarely publish substantial enterprise architecture descriptions. However, government agencies have begun to publish architectural descriptions they have developed. Examples include:

- US Department of the Interior

- US Department of Defense Business Enterprise Architecture, or the 2008 BEAv5.0 version

- Treasury Enterprise Architecture Framework

Relationship to Other Disciplines

According to the Federation of EA Professional Organizations (FEAPO), Enterprise Architecture interacts with a wide array of other disciplines commonly found in business settings. According to FEAPO:

> An Enterprise Architecture practice collaborates with many inter--connected disciplines including performance engineering and management, process engineering and management, IT and enterprise portfolio management, governance and compliance, IT strategiplanning, risk analysis, information management, metadata management, and a wide variety of technical disciplines as well as organizational disciplines such as organizational development, transformation, innovation, and learning. Increasingly, many practitioners have stressed the important relationship of Enterprise Architecture with emerging holistidesign practices such as design thinking, systems thinking, and user experience design.

As Enterprise Architecture has emerged in various organizations, the broad reach has resulted in this business role being included in information technology governance process in many organizations. While this may imply that enterprise architecture is closely tied to IT, it should be viewed in the broader context of business optimization in that it addresses business architecture, performance management and process architecture as well as more technical subjects.

Discussions of the intersection of Enterprise Architecture and various IT practices have been published by various IT analysis firms. Gartner and Forrester have stressed the important relationship of Enterprise Architecture with emerging holistidesign practices such as Design Thinking and User Experience Design. Analyst firm Real Story Group suggested that Enterprise Architecture and the emerging concept of the Digital workplace were "two sides to the same coin." The Cutter Consortium describes Enterprise Architecture as an information and knowledge-based discipline.

The enterprise architecture of an organization is too complex and extensive to document in its entirety, so knowledge management techniques provide a way to explore and analyze these hidden, tacit or implicit areas. In return, enterprise architecture provides a way of documenting the components of an organization and their interaction in a systemiand holistiway that complements knowledge management.

Enterprise Architecture has been discussed, in various venues, as having a relationship with Service Oriented Architecture, a particular style of application integration. Current research points to Enterprise Architecture as a key enabler to the success of efforts to use SOA as an enterprise-wide integration pattern.

Notable Enterprise Architecture Tools

The following table lists the most notable enterprise architecture tools as recognized by Gartner and Forrester Research in their most recent reports.

Product	Vendor	Headquarters	Latest stable release	Stable release date
ABACUS	Avolution	Australia	4.5	December 2015
ADOit	BOC Group	Austria	7.0	June 2016
BiZZdesign Architect	BiZZdesign	Netherlands	4.8.0	December 2015
ARIS	Software AG (formerly IDS Scheer)	Germany	9.0	March 2013
Casewise Suite/Evolve	Casewise	United Kingdom	2015/3.0	November 2015
Enterprise Architect	Sparx Systems	Australia	12	January 2015
iteraplan	iteratec	Germany	5.1	October 2015
leanIX	LeanIX	Germany	2.2	February 2016
Mega Suite	MEGA International Srl.	France	Release 3	August 2015
planningIT	Software AG (formerly alfabet)	Germany	8.0	November 2012
PowerDesigner	SAP-Sybase	Germany	16.0	November 2011
ProVision	OpenText (formerly Metastorm)	Canada	9.0	September 2012
QualiWare	QualiWare	Denmark	6.3	January 2016
SAMU	Atoll Technologies	Hungary	5.4	January 2016
QPR EnterpriseArchitect	QPR Software	Finland	2015.1	October 2015
System Architect	IBM (formerly Telelogic)	United States	11.4.3.6	December 2015
Troux	Troux Technologies (formerly Computas Technology)	United States	9.1.2	March 2013
Product	Vendor	Headquarters	Latest stable release	Stable release date

Criticism

Despite the benefits that enterprise architecture claims to provide, for more than a decade a number of industry leaders, writers, and leading organizations have raised concerns about enterprise architecture as an effective practice. Here is a partial list:

- In 2007, noted computer scientist Ivar Jacobson (a major contributor to UML and pioneer in OO software development) gave his assessment of enterprise architecture: "Around the world introducing an Enterprise Architecture EA has been an initiative for most financial institutions (banks, insurance companies, government, etc.) for the last five years or so, and it is not over. I have been working with such companies and helped some of them to avoid making the worst mistakes. Most EA initiatives failed. My guess is that more than 90% never really resulted in anything useful."

- In a 2007 report, on enterprise architecture, Gartner predicted that "... by 2012 40% of [2007's] enterprise architecture programs will be stopped."

- A 2008 study, by performed by Erasmus University Rotterdam and software company IDS Scheer concluded that two-thirds of enterprise architecture projects failed to improve business and IT alignment.

- In a 2009 article, industry commentator Dion Hinchcliffe wrote that traditional enterprise architecture might be "broken": "At its very best, enterprise architecture provides the bright lines that articulate the full range of possibilities for a business, even describing how to go about getting there. ... Recently there's a growing realization that traditional enterprise architecture as it's often practiced today might be broken in some important way. What might be wrong and how to fix it are the questions du jour."

- In 2011, federal enterprise architecture consultant Stanley Gaver released a report that examined problems within the United States federal government's enterprise architecture program. Mr. Gaver concluded that the federal enterprise architecture program had mostly failed; this conclusion was corroborated by a similar one made by the federal government at an October 2010 meeting that was held to determine why the federal enterprise architecture program was not "as influential and successful as in the past."

A key concern about EA has been the difficulty in arriving at metrics of success, because of the broad-brush and often opaque nature of EA projects.

References

- Clements, Paul; Felix Bachmann; Len Bass; David Garlan; James Ivers; Reed Little; Paulo Merson; Robert Nord; Judith Stafford (2010). Documenting Software Architectures: Views and Beyond, Second Edition. Boston: Addison-Wesley. ISBN 0-321-55268-7.

- Bass, Len; Paul Clements; Rick Kazman (2012). Software Architecture In Practice, Third Edition. Boston: Addison-Wesley. ISBN 978-0-321-81573-6.

- Jansen, A.; Bosch, J. (2005). "Software Architecture as a Set of Architectural Design Decisions". 5th Working IEEE/IFIP Conference on Software Architecture (WICSA'05). p. 109. doi:10.1109/WICSA.2005.61. ISBN 0-7695-2548-2.

- Ali Babar, Muhammad; Dingsoyr, Torgeir; Lago, Patricia; van Vliet, Hans (2009). Software Architecture Knowledge Management. Dordrecht Heidelberg London New York: Springer. ISBN 978-3-642-02373-6.

- Angelov, Samuil; Grefen, Paul; Greefhorst, Danny. "A Classification of Software Reference Architectures: Analyzing Their Success and Effectiveness". Proc. of WICSA/ECSA 2009. IEEE: 141–150. doi:10.1109/ WICSA.2009.5290800. Retrieved 13 November 2015.

- Obbink, H.; Kruchten, P.; Kozaczynski, W.; Postema, H.; Ran, A.; Dominick, L.; Kazman, R.; Hilliard, R.; Tracz, W.; Kahane, E. (Feb 6, 2002). "Software Architecture Review and Assessment (SARA) Report" (PDF). Retrieved November 1, 2015.

- Rosa and Sampaio. "SOA Governance Through Enterprise Architecture". Oracle.com. Oracle. Retrieved December 19, 2014.

- Poort, Eltjo; van Vliet, Hans (September 2012). "RCDA: Architecting as a risk- and cost management discipline".

The Journal of Systems and Software. Elsevier. 85 (9): 1995–2013. doi:10.1016/j.jss.2012.03.071.

- UCI Software Architecture Research – UCI Software Architecture Research: Architectural Styles. Isr.uci.edu. Retrieved on 2013-07-21.

- Federation of EA Professional Organizations, Common Perspectives on Enterprise Architecture, Architecture and Governance Magazine, Issue 9-4, November 2013 (2013). Retrieved on November 19, 2013.

- ISO/IEC/IEEE (2011). "ISO/IEC/IEEE 42010:2011 Systems and software engineering – Architecture description". Retrieved 2012-09-12.

- ISO/IEC (2011). "ISO/IEC 25010:2011 Systems and software engineering – Systems and software Quality Requirements and Evaluation (SQuaRE) – System and software quality models". Retrieved 2012-10-08.

Applications of Control Systems

The following chapter elucidates the applications that are related to control system. It discusses the functions of control systems in a critical manner providing key analysis to the subject matter. The applications explained are electrical network, digital signal processor, microcontroller and cruise control.

Electrical Network

An electrical network is an interconnection of electrical components (e.g. batteries, resistors, inductors, capacitors, switches) or a model of such an interconnection, consisting of electrical elements (e.g. voltage sources, current sources, resistances, inductances, capacitances). An electrical circuit is a network consisting of a closed loop, giving a return path for the current. Linear electrical networks, a special type consisting only of sources (voltage or current), linear lumped elements (resistors, capacitors, inductors), and linear distributed elements (transmission lines), have the property that signals are linearly superimposable. They are thus more easily analyzed, using powerful frequency domain methods such as Laplace transforms, to determine DC response, AC response, and transient response.

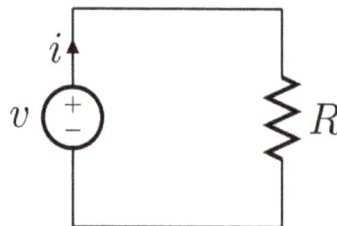

A simple electricircuit made up of a voltage source and a resistor. Here, , according to Ohm's law.

A resistive circuit is a circuit containing only resistors and ideal current and voltage sources. Analysis of resistive circuits is less complicated than analysis of circuits containing capacitors and inductors. If the sources are constant (DC) sources, the result is a DC circuit.

A network that contains active electronicomponents is known as an *electronicircuit*. Such networks are generally nonlinear and require more complex design and analysis tools.

Classification

By passivity

An active network is a network that contains an active source - either a voltage source or current source.

A passive network is a network that does not contain an active source.

By linearity

A network is linear if its signals obey the principle of superposition; otherwise it is non-linear.

Classification of Sources

Sources can be classified as independent sources and dependent sources

Independent

Ideal Independent Source maintains same voltage or current regardless of the other elements present in the circuit. Its value is either constant (DC) or sinusoidal (AC). The strength of voltage or current is not changed by any variation in connected network.

Dependent

Dependent Sources depend upon a particular element of the circuit for delivering the power or voltage or current depending upon the type of source it is.

Electrical Laws

A number of electrical laws apply to all electrical networks. These include:

- Kirchhoff's current law: The sum of all currents entering a node is equal to the sum of all currents leaving the node.

- Kirchhoff's voltage law: The directed sum of the electrical potential differences around a loop must be zero.

- Ohm's law: The voltage across a resistor is equal to the product of the resistance and the current flowing through it.

- Norton's theorem: Any network of voltage or current sources and resistors is electrically equivalent to an ideal current source in parallel with a single resistor.

- Thévenin's theorem: Any network of voltage or current sources and resistors is electrically equivalent to a single voltage source in series with a single resistor.

- superposition theorem: In a linear network with several independent sources, the response in a particular branch when all the sources are acting simultaneously is equal to the linear sum of individual responses calculated by taking one independent source at a time.

Other more complex laws may be needed if the network contains nonlinear or reactive components. Non-linear self-regenerative heterodyning systems can be approximated. Applying these laws results in a set of simultaneous equations that can be solved either algebraically or numerically.

Design Methods

To design any electrical circuit, either analog or digital, electrical engineers need to be able to predict the voltages and currents at all places within the circuit. Simple linear circuits can be analyzed by hand using complex number theory. In more complex cases the circuit may be analyzed with specialized computer programs or estimation techniques such as the piecewise-linear model.

Circuit simulation software, such as HSPICE (an analog circuit simulator), and languages such as VHDL-AMS and verilog-AMS allow engineers to design circuits without the time, cost and risk of error involved in building circuit prototypes.

Network Simulation Software

More complex circuits can be analyzed numerically with software such as SPICE or GNUCAP, or symbolically using software such as SapWin.

Linearization Around Operating Point

When faced with a new circuit, the software first tries to find a steady state solution, that is, one where all nodes conform to Kirchhoff's current law *and* the voltages across and through each element of the circuit conform to the voltage/current equations governing that element.

Once the steady state solution is found, the operating points of each element in the circuit are known. For a small signal analysis, every non-linear element can be linearized around its operation point to obtain the small-signal estimate of the voltages and currents. This is an application of Ohm's Law. The resulting linear circuit matrix can be solved with Gaussian elimination.

Piecewise-linear Approximation

Software such as the PLECS interface to Simulink uses piecewise-linear approximation of the equations governing the elements of a circuit. The circuit is treated as a completely linear network of ideal diodes. Every time a diode switches from on to off or vice versa, the configuration of the linear network changes. Adding more detail to the approximation of equations increases the accuracy of the simulation, but also increases its running time.

Digital Signal Processor

A digital signal processor chip found in a guitar effects unit.

A digital signal processor (DSP) is a specialized microprocessor (or a SIP block), with its architecture optimized for the operational needs of digital signal processing.

The goal of DSPs is usually to measure, filter and/or compress continuous real-world analog signals. Most general-purpose microprocessors can also execute digital signal processing algorithms successfully, but dedicated DSPs usually have better power efficiency thus they are more suitable in portable devices such as mobile phones because of power consumption constraints. DSPs often use special memory architectures that are able to fetch multiple data and/or instructions at the same time.

Overview

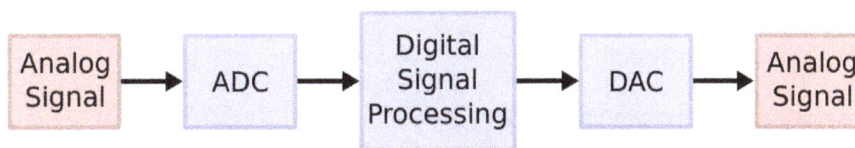

A typical digital processing system

Digital signal processing algorithms typically require a large number of mathematical operations to be performed quickly and repeatedly on a series of data samples. Signals (perhaps from audio or video sensors) are constantly converted from analog to digital, manipulated digitally, and then converted back to analog form. Many DSP applications have constraints on latency; that is, for the system to work, the DSP operation must be completed within some fixed time, and deferred (or batch) processing is not viable.

Most general-purpose microprocessors and operating systems can execute DSP algorithms successfully, but are not suitable for use in portable devices such as mobile phones and PDAs because of power efficiency constraints. A specialized digital signal processor, however, will tend to provide a lower-cost solution, with better performance, lower latency, and no requirements for specialized cooling or large batteries.

The architecture of a digital signal processor is optimized specifically for digital signal processing. Most also support some of the features as an applications processor or microcontroller, since signal processing is rarely the only task of a system. Some useful features for optimizing DSP algorithms are outlined below.

Architecture

Software Architecture

By the standards of general-purpose processors, DSP instruction sets are often highly irregular; while traditional instruction sets are made up of more general instructions that allow them to perform a wider variety of operations, instruction sets optimized for digital signal processing contain instructions for common mathematical operations that occur frequently in DSP calculations. Both traditional and DSP-optimized instruction sets are able to compute any arbitrary operation but an operation that might require multiple ARM or x86 instructions to compute might require only one instruction in a DSP optimized instruction set.

One implication for software architecture is that hand-optimized assembly-code routines are commonly packaged into libraries for re-use, instead of relying on advanced compiler technologies to handle essential algorithms. Even with modern compiler optimizations hand-optimized assembly code is more efficient and many common algorithms involved in DSP calculations are hand-written in order to take full advantage of the architectural optimizations.

Instruction Sets

- multiply–accumulates (MACs, including fused multiply–add, FMA) operations
  - used extensively in all kinds of matrix operations
    - convolution for filtering
    - dot product
    - polynomial evaluation
  - Fundamental DSP algorithms depend heavily on multiply–accumulate performance
    - FIR filters
    - Fast Fourier transform (FFT)
- Instructions to increase parallelism:
  - SIMD
  - VLIW
  - superscalar architecture
- Specialized instructions for modulo addressing in ring buffers and bit-reversed addressing mode for FFT cross-referencing
- Digital signal processors sometimes use time-stationary encoding to simplify hardware and increase coding efficiency.
- Multiple arithmetiunits may require memory architectures to support several accesses per instruction cycle
- Special loop controls, such as architectural support for executing a few instruction words in a very tight loop without overhead for instruction fetches or exit testing

Data instructions

- Saturation arithmetic, in which operations that produce overflows will accumulate at the maximum (or minimum) values that the register can hold rather than wrapping around (maximum+1 doesn't overflow to minimum as in many general-purpose CPUs, instead it stays at maximum). Sometimes various sticky bits operation modes are available.
- Fixed-point arithmetiis often used to speed up arithmetiprocessing

- Single-cycle operations to increase the benefits of pipelining

Program flow

- Floating-point unit integrated directly into the datapath

- Pipelined architecture

- Highly parallel multiplier–accumulators (MAC units)

- Hardware-controlled looping, to reduce or eliminate the overhead required for looping operations

Hardware Architecture

Memory Architecture

DSPs are usually optimized for streaming data and use special memory architectures that are able to fetch multiple data and/or instructions at the same time, such as the Harvard architecture or Modified von Neumann architecture, which use separate program and data memories (sometimes even concurrent access on multiple data buses).

DSPs can sometimes rely on supporting code to know about cache hierarchies and the associated delays. This is a tradeoff that allows for better performance. In addition, extensive use of DMA is employed.

Addressing and Virtual Memory

DSPs frequently use multi-tasking operating systems, but have no support for virtual memory or memory protection. Operating systems that use virtual memory require more time for context switching among processes, which increases latency.

- Hardware modulo addressing

    o Allows circular buffers to be implemented without having to test for wrapping

- Bit-reversed addressing, a special addressing mode

    o useful for calculating FFTs

- Exclusion of a memory management unit

- Memory-address calculation unit

History

Prior to the advent of stand-alone DSP chips discussed below, most DSP applications were implemented using bit-slice processors. The AMD 2901 bit-slice chip with its family of components was a very popular choice. There were reference designs from AMD, but very often the specifics of a particular design were application specific. These bit slice architectures would sometimes include a peripheral multiplier chip. Examples of these multipliers were a series from TRW including the

TDC1008 and TDC1010, some of which included an accumulator, providing the requisite multiply–accumulate (MAC) function.

In 1976, Richard Wiggins proposed the Speak & Spell concept to Paul Breedlove, Larry Brantingham, and Gene Frantz at Texas Instrument's Dallas research facility. Two years later in 1978 they produced the first Speak & Spell, with the technological centerpiece being the TMS5100, the industry's first digital signal processor. It also set other milestones, being the first chip to use Linear predictive coding to perform speech synthesis.

In 1978, Intel released the 2920 as an "analog signal processor". It had an on-chip ADC/DAC with an internal signal processor, but it didn't have a hardware multiplier and was not successful in the market. In 1979, AMI released the S2811. It was designed as a microprocessor peripheral, and it had to be initialized by the host. The S2811 was likewise not successful in the market.

In 1980 the first stand-alone, complete DSPs – the NEC µPD7720 and AT&T DSP1 – were presented at the International Solid-State Circuits Conference '80. Both processors were inspired by the research in PSTN telecommunications.

The Altamira DX-1 was another early DSP, utilizing quad integer pipelines with delayed branches and branch prediction.

Another DSP produced by Texas Instruments (TI), the TMS32010 presented in 1983, proved to be an even bigger success. It was based on the Harvard architecture, and so had separate instruction and data memory. It already had a special instruction set, with instructions like load-and-accumulate or multiply-and-accumulate. It could work on 16-bit numbers and needed 390 ns for a multiply–add operation. TI is now the market leader in general-purpose DSPs.

About five years later, the second generation of DSPs began to spread. They had 3 memories for storing two operands simultaneously and included hardware to accelerate tight loops, they also had an addressing unit capable of loop-addressing. Some of them operated on 24-bit variables and a typical model only required about 21 ns for a MAC. Members of this generation were for example the AT&T DSP16A or the Motorola 56000.

The main improvement in the third generation was the appearance of application-specifiunits and instructions in the data path, or sometimes as coprocessors. These units allowed direct hardware acceleration of very specifibut complex mathematical problems, like the Fourier-transform or matrix operations. Some chips, like the Motorola MC68356, even included more than one processor core to work in parallel. Other DSPs from 1995 are the TI TMS320C541 or the TMS 320C80.

The fourth generation is best characterized by the changes in the instruction set and the instruction encoding/decoding. SIMD extensions were added, VLIW and the superscalar architecture appeared. As always, the clock-speeds have increased, a 3 ns MAC now became possible.

Modern DSPs

Modern signal processors yield greater performance; this is due in part to both technological and architectural advancements like lower design rules, fast-access two-level cache, (E)DMA circuitry and a wider bus system. Not all DSPs provide the same speed and many kinds of signal processors exist, each one of them being better suited for a specifitask, ranging in price from about US$1.50 to US$300

Texas Instruments produces the C6000 series DSPs, which have clock speeds of 1.2 GHz and implement separate instruction and data caches. They also have an 8 MiB 2nd level cache and 64 EDMA channels. The top models are capable of as many as 8000 MIPS (instructions per second), use VLIW (very long instruction word), perform eight operations per clock-cycle and are compatible with a broad range of external peripherals and various buses (PCI/serial/etc). TMS320C6474 chips each have three such DSPs, and the newest generation C6000 chips support floating point as well as fixed point processing.

Freescale produces a multi-core DSP family, the MSC81xx. The MSC81xx is based on StarCore Architecture processors and the latest MSC8144 DSP combines four programmable SC3400 StarCore DSP cores. Each SC3400 StarCore DSP core has a clock speed of 1 GHz.

XMOS produces a multi-core multi-threaded line of processor well suited to DSP operations, They come in various speeds ranging from 400 to 1600 MIPS. The processors have a multi-threaded architecture that allows up to 8 real-time threads per core, meaning that a 4 core device would support up to 32 real time threads. Threads communicate between each other with buffered channels that are capable of up to 80 Mbit/s. The devices are easily programmable in C and aim at bridging the gap between conventional micro-controllers and FPGAs

CEVA, Inc. produces and licenses three distinct families of DSPs. Perhaps the best known and most widely deployed is the CEVA-TeakLite DSP family, a classimemory-based architecture, with 16-bit or 32-bit word-widths and single or dual MACs. The CEVA-X DSP family offers a combination of VLIW and SIMD architectures, with different members of the family offering dual or quad 16-bit MACs. The CEVA-XC DSP family targets Software-defined Radio (SDR) modem designs and leverages a unique combination of VLIW and Vector architectures with 32 16-bit MACs.

Analog Devices produce the SHARC-based DSP and range in performance from 66 MHz/198 MFLOPS (million floating-point operations per second) to 400 MHz/2400 MFLOPS. Some models support multiple multipliers and ALUs, SIMD instructions and audio processing-specificomponents and peripherals. The Blackfin family of embedded digital signal processors combine the features of a DSP with those of a general use processor. As a result, these processors can run simple operating systems like μCLinux, velOSity and Nucleus RTOS while operating on real-time data.

NXP Semiconductors produce DSPs based on TriMedia VLIW technology, optimized for audio and video processing. In some products the DSP core is hidden as a fixed-function block into a SoC, but NXP also provides a range of flexible single core media processors. The TriMedia media processors support both fixed-point arithmetias well as floating-point arithmetic, and have specifiinstructions to deal with complex filters and entropy coding.

CSR produces the Quatro family of SoCs that contain one or more custom Imaging DSPs optimized for processing document image data for scanner and copier applications.

Most DSPs use fixed-point arithmetic, because in real world signal processing the additional range provided by floating point is not needed, and there is a large speed benefit and cost benefit due to reduced hardware complexity. Floating point DSPs may be invaluable in applications where a wide dynamirange is required. Product developers might also use floating point DSPs to reduce the cost and complexity of software development in exchange for more expensive hardware, since it is generally easier to implement algorithms in floating point.

Generally, DSPs are dedicated integrated circuits; however DSP functionality can also be produced by using field-programmable gate array chips (FPGAs).

Embedded general-purpose RISC processors are becoming increasingly DSP like in functionality. For example, the OMAP3 processors include a ARM Cortex-A8 and C6000 DSP.

In Communications a new breed of DSPs offering the fusion of both DSP functions and H/W acceleration function is making its way into the mainstream. Such Modem processors include ASOCS ModemX and CEVA's XC4000.

Microcontroller

The die from an Intel 8742, an 8-bit microcontroller that includes a CPU running at 12 MHz, 128 bytes of RAM, 2048 bytes of EPROM, and I/O in the same chip.

Two ATmega microcontrollers

A microcontroller (or MCU, short for *microcontroller unit*) is a small computer (SoC) on a single integrated circuit containing a processor core, memory, and programmable input/output peripherals. Program memory in the form of FerroelectriRAM, NOR flash or OTP ROM is also often included on chip, as well as a typically small amount of RAM. Microcontrollers are designed for embedded applications, in contrast to the microprocessors used in personal computers or other general purpose applications consisting of various discrete chips.

Microcontrollers are used in automatically controlled products and devices, such as automobile engine control systems, implantable medical devices, remote controls, office machines, appliances, power tools, toys and other embedded systems. By reducing the size and cost compared to a design that uses a separate microprocessor, memory, and input/output devices, microcontrollers make it economical to digitally control even more devices and processes. Mixed signal microcontrollers are common, integrating analog components needed to control non-digital electronisystems.

Some microcontrollers may use four-bit words and operate at frequencies as low as 4 kHz, for low power consumption (single-digit milliwatts or microwatts). They will generally have the ability to retain functionality while waiting for an event such as a button press or other interrupt; power consumption while sleeping (CPU clock and most peripherals off) may be just nanowatts, making many of them well suited for long lasting battery applications. Other microcontrollers may serve performance-critical roles, where they may need to act more like a digital signal processor (DSP), with higher clock speeds and power consumption.

History

The first microprocessor was the 4-bit Intel 4004 released in 1971, with the Intel 8008 and other more capable microprocessors becoming available over the next several years. However, both processors required external chips to implement a working system, raising total system cost, and making it impossible to economically computerize appliances.

The Smithsonian Institution credits TI engineers Gary Boone and Michael Cochran with the successful creation of the first microcontroller in 1971. The result of their work was the TMS 1000, which became commercially available in 1974. It combined read-only memory, read/write memory, processor and clock on one chip and was targeted at embedded systems.

Partly in response to the existence of the single-chip TMS 1000, Intel developed a computer system on a chip optimized for control applications, the Intel 8048, with commercial parts first shipping in 1977. It combined RAM and ROM on the same chip. This chip would find its way into over one billion PC keyboards, and other numerous applications. At that time Intel's President, Luke J. Valenter, stated that the microcontroller was one of the most successful in the company's history, and expanded the division's budget over 25%.

Most microcontrollers at this time had concurrent variants. One had an erasable EPROM program memory, with a transparent quartz window in the lid of the package to allow it to be erased by exposure to ultraviolet light, often used for prototyping. The other was either a mask programmed ROM from the manufacturer for large series, or a PROM variant which was only programmable once; sometimes this was signified with the designation OTP, standing for "one-time programmable". The PROM was of identical type of memory as the EPROM, but because there was no way to expose it to ultraviolet light, it could not be erased. The erasable versions required ceramipackages with quartz windows, making them significantly more expensive than the OTP versions, which could be made in lower-cost opaque plastipackages. For the erasable variants, quartz was required, instead of less expensive glass, for its transparency to ultraviolet—glass is largely opaque to UV— but the main cost differentiator was the ceramipackage itself.

In 1993, the introduction of EEPROM memory allowed microcontrollers (beginning with the Microchip PIC16x84) to be electrically erased quickly without an expensive package as required for EPROM, allowing both rapid prototyping, and In System Programming. (EEPROM technology had been available prior to this time, but the earlier EEPROM was more expensive and less durable, making it unsuitable for low-cost mass-produced microcontrollers.) The same year, Atmel introduced the first microcontroller using Flash memory, a special type of EEPROM. Other companies rapidly followed suit, with both memory types.

Cost has plummeted over time, with the cheapest 8-bit microcontrollers being available for under 0.25 USD in quantity (thousands) in 2009, and some 32-bit microcontrollers around US$1 for similar quantities.

Nowadays microcontrollers are cheap and readily available for hobbyists, with large online communities around certain processors.

In the future, MRAM could potentially be used in microcontrollers as it has infinite endurance and its incremental semiconductor wafer process cost is relatively low.

Volumes

In 2002, about 55% of all CPUs sold in the world were 8-bit microcontrollers and microprocessors. Over two billion 8-bit microcontrollers were sold in 1997, and according to Semico, over four billion 8-bit microcontrollers were sold in 2006. More recently, Semico has claimed the MCU market grew 36.5% in 2010 and 12% in 2011.

A typical home in a developed country is likely to have only four general-purpose microprocessors but around three dozen microcontrollers. A typical mid-range automobile has as many as 30 or more microcontrollers. They can also be found in many electrical devices such as washing machines, microwave ovens, and telephones.

Historically, the 8-bit segment has dominated the MCU market [..] 16-bit microcontrollers became the largest volume MCU category in 2011, overtaking 8-bit devices for the first time that year [..] IC Insights believes the makeup of the MCU market will undergo substantial changes in the next five years with 32-bit devices steadily grabbing a greater share of sales and unit volumes. By 2017, 32-bit MCUs are expected to account for 55% of microcontroller sales [..] In terms of unit volumes, 32-bit MCUs are expected account for 38% of microcontroller shipments in 2017, while 16-bit devices will represent 34% of the total, and 4-/8-bit designs are forecast to be 28% of units sold that year. The 32-bit MCU market is expected to grow rapidly due to increasing demand for higher levels of precision in embedded-processing systems and the growth in connectivity using the Internet. [..] In the next few years, complex 32-bit MCUs are expected to account for over 25% of the processing power in vehicles.

— IC Insights, MCU Market on Migration Path to 32-bit and ARM-based Devices

In 2012, following a global crisis – a worst ever annual sales decline and recovery and average sales price year-over-year plunging 17% – the biggest reduction since the 1980s, the average price for a microcontroller was US$0.88 ($0.69 for 4-/8-bit, $0.59 for 16-bit, $1.76 for 32-bit).

In 2012, worldwide sales of 8-bit microcontrollers were around $4 billion because they were so useful that many companies needed them to be able to progress into better technology. In 2012, 4-bit microcontrollers also see significant sales.

In 2015, 8-bit microcontrollers can be bought for \$0.311 (1,000 units), 16-bit for \$0.385 (1,000 units), and 32-bit for \$0.378 (1,000 units but at \$0.35 for 5,000).

A PIC 18F8720 microcontroller in an 80-pin TQFP package.

Embedded Design

A microcontroller can be considered a self-contained system with a processor, memory and peripherals and can be used as an embedded system. The majority of microcontrollers in use today are embedded in other machinery, such as automobiles, telephones, appliances, and peripherals for computer systems.

While some embedded systems are very sophisticated, many have minimal requirements for memory and program length, with no operating system, and low software complexity. Typical input and output devices include switches, relays, solenoids, LEDs, small or custom liquid-crystal displays, radio frequency devices, and sensors for data such as temperature, humidity, light level etc. Embedded systems usually have no keyboard, screen, disks, printers, or other recognizable I/O devices of a personal computer, and may lack human interaction devices of any kind.

Interrupts

Microcontrollers must provide real-time (predictable, though not necessarily fast) response to events in the embedded system they are controlling. When certain events occur, an interrupt system can signal the processor to suspend processing the current instruction sequence and to begin an interrupt service routine (ISR, or "interrupt handler") which will perform any processing required based on the source of the interrupt, before returning to the original instruction sequence. Possible interrupt sources are device dependent, and often include events such as an internal timer overflow, completing an analog to digital conversion, a logilevel change on an input such as from a button being pressed, and data received on a communication link. Where power consumption is important as in batteried devices, interrupts may also wake a microcontroller from a low-power sleep state where the processor is halted until required to do something by a peripheral event.

Programs

Typically microcontroller programs must fit in the available on-chip memory, since it would be costly to provide a system with external, expandable memory. Compilers and assemblers are used to convert both high-level and assembly language codes into a compact machine code for storage in the microcontroller's memory. Depending on the device, the program memory may be permanent, read-only memory that can only be programmed at the factory, or it may be field-alterable flash or erasable read-only memory.

Manufacturers have often produced special versions of their microcontrollers in order to help the hardware and software development of the target system. Originally these included EPROM versions that have a "window" on the top of the device through which program memory can be erased by ultraviolet light, ready for reprogramming after a programming ("burn") and test cycle. Since 1998, EPROM versions are rare and have been replaced by EEPROM and flash, which are easier to use (can be erased electronically) and cheaper to manufacture.

Other versions may be available where the ROM is accessed as an external device rather than as internal memory, however these are becoming rare due to the widespread availability of cheap microcontroller programmers.

The use of field-programmable devices on a microcontroller may allow field update of the firmware or permit late factory revisions to products that have been assembled but not yet shipped. Programmable memory also reduces the lead time required for deployment of a new product.

Where hundreds of thousands of identical devices are required, using parts programmed at the time of manufacture can be economical. These "mask programmed" parts have the program laid down in the same way as the logiof the chip, at the same time.

A customizable microcontroller incorporates a block of digital logithat can be personalized for additional processing capability, peripherals and interfaces that are adapted to the requirements of the application. One example is the AT91CAP from Atmel.

Other Microcontroller Features

Microcontrollers usually contain from several to dozens of general purpose input/output pins (GPIO). GPIO pins are software configurable to either an input or an output state. When GPIO pins are configured to an input state, they are often used to read sensors or external signals. Configured to the output state, GPIO pins can drive external devices such as LEDs or motors, often indirectly, through external power electronics.

Many embedded systems need to read sensors that produce analog signals. This is the purpose of the analog-to-digital converter (ADC). Since processors are built to interpret and process digital data, i.e. 1s and 0s, they are not able to do anything with the analog signals that may be sent to it by a device. So the analog to digital converter is used to convert the incoming data into a form that the processor can recognize. A less common feature on some microcontrollers is a digital-to-analog converter (DAC) that allows the processor to output analog signals or voltage levels.

In addition to the converters, many embedded microprocessors include a variety of timers as well.

One of the most common types of timers is the Programmable Interval Timer (PIT). A PIT may either count down from some value to zero, or up to the capacity of the count register, overflowing to zero. Once it reaches zero, it sends an interrupt to the processor indicating that it has finished counting. This is useful for devices such as thermostats, which periodically test the temperature around them to see if they need to turn the air conditioner on, the heater on, etc.

A dedicated Pulse Width Modulation (PWM) block makes it possible for the CPU to control power converters, resistive loads, motors, etc., without using lots of CPU resources in tight timer loops.

Universal Asynchronous Receiver/Transmitter (UART) block makes it possible to receive and transmit data over a serial line with very little load on the CPU. Dedicated on-chip hardware also often includes capabilities to communicate with other devices (chips) in digital formats such as Inter-Integrated Circuit (I²C), Serial Peripheral Interface (SPI), Universal Serial Bus (USB), and Ethernet.

Higher Integration

Die of a PIC12C508 8-bit, fully static, EEPROM/EPROM/ROM-based CMOS microcontroller manufactured by Microchip Technology using a 1200 nanometre process.

Die of a STM32F100C4T6B ARM Cortex-M3 microcontroller with 16 kilobytes flash memory, 24 MHz Central Processing Unit (CPU), motor control and Consumer Electronics Control (CEC) functions. Manufactured by STMicroelectronics.

Micro-controllers may not implement an external address or data bus as they integrate RAM and non-volatile memory on the same chip as the CPU. Using fewer pins, the chip can be placed in a much smaller, cheaper package.

Integrating the memory and other peripherals on a single chip and testing them as a unit increases the cost of that chip, but often results in decreased net cost of the embedded system as a whole. Even if the cost of a CPU that has integrated peripherals is slightly more than the cost of a CPU and external peripherals, having fewer chips typically allows a smaller and cheaper circuit board, and reduces the labor required to assemble and test the circuit board, in addition to tending to decrease the defect rate for the finished assembly.

A micro-controller is a single integrated circuit, commonly with the following features:

- central processing unit - ranging from small and simple 4-bit processors to complex 32-bit or 64-bit processors

- volatile memory (RAM) for data storage

- ROM, EPROM, EEPROM or Flash memory for program and operating parameter storage

- discrete input and output bits, allowing control or detection of the logistate of an individual package pin

- serial input/output such as serial ports (UARTs)

- other serial communications interfaces like I²C, Serial Peripheral Interface and Controller Area Network for system interconnect

- peripherals such as timers, event counters, PWM generators, and watchdog

- clock generator - often an oscillator for a quartz timing crystal, resonator or RC circuit

- many include analog-to-digital converters, some include digital-to-analog converters

- in-circuit programming and in-circuit debugging support

This integration drastically reduces the number of chips and the amount of wiring and circuit board space that would be needed to produce equivalent systems using separate chips. Furthermore, on low pin count devices in particular, each pin may interface to several internal peripherals, with the pin function selected by software. This allows a part to be used in a wider variety of applications than if pins had dedicated functions.

Micro-controllers have proved to be highly popular in embedded systems since their introduction in the 1970s.

Some microcontrollers use a Harvard architecture: separate memory buses for instructions and data, allowing accesses to take place concurrently. Where a Harvard architecture is used, instruction words for the processor may be a different bit size than the length of internal memory and registers; for example: 12-bit instructions used with 8-bit data registers.

The decision of which peripheral to integrate is often difficult. The microcontroller vendors often trade operating frequencies and system design flexibility against time-to-market requirements from their customers and overall lower system cost. Manufacturers have to balance the need to minimize the chip size against additional functionality.

Microcontroller architectures vary widely. Some designs include general-purpose microprocessor cores, with one or more ROM, RAM, or I/O functions integrated onto the package. Other designs are purpose built for control applications. A micro-controller instruction set usually has many instructions intended for bit manipulation (bit-wise operations) to make control programs more compact. For example, a general purpose processor might require several instructions to test a bit in a register and branch if the bit is set, where a micro-controller could have a single instruction to provide that commonly required function.

Microcontrollers typically do not have a math coprocessor, so floating point arithmetiis performed by software.

Programming Environments

Microcontrollers were originally programmed only in assembly language, but various high-level programming languages, such as C, Python and JavaScript, are now also in common use to target microcontrollers and embedded systems. These languages are either designed specially for the purpose, or versions of general purpose languages such as the C programming language. Compilers for general purpose languages will typically have some restrictions as well as enhancements to better support the unique characteristics of microcontrollers. Some microcontrollers have environments to aid developing certain types of applications. Microcontroller vendors often make tools freely available to make it easier to adopt their hardware.

Many microcontrollers are so quirky that they effectively require their own non-standard dialects of C, such as SDCC for the 8051, which prevent using standard tools (such as code libraries or stationalysis tools) even for code unrelated to hardware features. Interpreters are often used to hide such low level quirks.

Interpreter firmware is also available for some microcontrollers. For example, BASIC on the early microcontrollers Intel 8052; BASIC and FORTH on the Zilog Z8 as well as some modern devices. Typically these interpreters support interactive programming.

Simulators are available for some microcontrollers. These allow a developer to analyze what the behavior of the microcontroller and their program should be if they were using the actual part. A simulator will show the internal processor state and also that of the outputs, as well as allowing input signals to be generated. While on the one hand most simulators will be limited from being unable to simulate much other hardware in a system, they can exercise conditions that may otherwise be hard to reproduce at will in the physical implementation, and can be the quickest way to debug and analyze problems.

Recent microcontrollers are often integrated with on-chip debug circuitry that when accessed by an in-circuit emulator via JTAG, allow debugging of the firmware with a debugger. A real-time ICE may allow viewing and/or manipulating of internal states while running. A tracing ICE can record executed program and MCU states before/after a trigger point.

Types of microcontrollers

As of 2008, there are several dozen microcontroller architectures and vendors including:

- ARM core processors (many vendors)

  o ARM Cortex-M cores are specifically targeted towards microcontroller applications

- Atmel AVR (8-bit), AVR32 (32-bit), and AT91SAM (32-bit)

- Cypress Semiconductor's M8C Core used in their PSoC (Programmable System-on-Chip)

- Freescale ColdFire (32-bit) and S08 (8-bit)

- Freescale 68HC11 (8-bit), and others based on the Motorola 6800 family

- Intel 8051, also manufactured by NXP Semiconductors, Infineon and many others

- Infineon: 8-bit XC800, 16-bit XE166, 32-bit XMC4000 (ARM based Cortex M4F), 32-bit TriCore and, 32-bit Aurix Tricore Bit microcontrollers

- MIPS

- Microchip Technology PIC, (8-bit PIC16, PIC18, 16-bit dsPIC33 / PIC24), (32-bit PIC32)

- NXP Semiconductors LPC1000, LPC2000, LPC3000, LPC4000 (32-bit), LPC900, LPC700 (8-bit)

- Parallax Propeller

- PowerPC ISE

- Rabbit 2000 (8-bit)

- Renesas Electronics: RL78 16-bit MCU; RX 32-bit MCU; SuperH; V850 32-bit MCU; H8; R8C 16-bit MCU

- Silicon Laboratories Pipelined 8-bit 8051 Microcontrollers and mixed-signal ARM-based 32-bit microcontrollers

- STMicroelectronics STM8 (8-bit), ST10 (16-bit) and STM32 (32-bit)

- Texas Instruments TI MSP430 (16-bit), MSP432 (32-bit), C2000 (32-bit)

- Toshiba TLCS-870 (8-bit/16-bit)

Many others exist, some of which are used in very narrow range of applications or are more like applications processors than microcontrollers. The microcontroller market is extremely fragmented, with numerous vendors, technologies, and markets. Note that many vendors sell or have sold multiple architectures.

Interrupt Latency

In contrast to general-purpose computers, microcontrollers used in embedded systems often seek to optimize interrupt latency over instruction throughput. Issues include both reducing the latency, and making it be more predictable (to support real-time control).

When an electronidevice causes an interrupt, during the context switch the intermediate results (registers) have to be saved before the software responsible for handling the interrupt can run. They must also be restored after that interrupt handler is finished. If there are more processor registers, this saving and restoring process takes more time, increasing the latency. Ways to reduce such context/restore latency include having relatively few registers in their central processing units (undesirable because it slows down most non-interrupt processing substantially), or at least having the hardware not save them all (this fails if the software then needs to compensate by saving the rest "manually"). Another technique involves spending silicon gates on "shadow registers": One or more duplicate registers used only by the interrupt software, perhaps supporting a dedicated stack.

Other factors affecting interrupt latency include:

- Cycles needed to complete current CPU activities. To minimize those costs, microcontrollers tend to have short pipelines (often three instructions or less), small write buffers, and ensure that longer instructions are continuable or restartable. RISC design principles ensure that most instructions take the same number of cycles, helping avoid the need for most such continuation/restart logic.

- The length of any critical section that needs to be interrupted. Entry to a critical section restricts concurrent data structure access. When a data structure must be accessed by an interrupt handler, the critical section must block that interrupt. Accordingly, interrupt latency is increased by however long that interrupt is blocked. When there are hard external constraints on system latency, developers often need tools to measure interrupt latencies and track down which critical sections cause slowdowns.

    o One common technique just blocks all interrupts for the duration of the critical section. This is easy to implement, but sometimes critical sections get uncomfortably long.

    o A more complex technique just blocks the interrupts that may trigger access to that data structure. This is often based on interrupt priorities, which tend to not correspond well to the relevant system data structures. Accordingly, this technique is used mostly in very constrained environments.

    o Processors may have hardware support for some critical sections. Examples include supporting atomiaccess to bits or bytes within a word, or other atomiaccess primitives like the LDREX/STREX exclusive access primitives introduced in the ARMv6 architecture.

- Interrupt nesting. Some microcontrollers allow higher priority interrupts to interrupt lower priority ones. This allows software to manage latency by giving time-critical interrupts higher priority (and thus lower and more predictable latency) than less-critical ones.

- Trigger rate. When interrupts occur back-to-back, microcontrollers may avoid an extra context save/restore cycle by a form of tail call optimization.

Lower end microcontrollers tend to support fewer interrupt latency controls than higher end ones.

Microcontroller Embedded Memory Technology

Since the emergence of microcontrollers, many different memory technologies have been used. Almost all microcontrollers have at least two different kinds of memory, a non-volatile memory for storing firmware and a read-write memory for temporary data.

Data

From the earliest microcontrollers to today, six-transistor SRAM is almost always used as the read/write working memory, with a few more transistors per bit used in the register file. FRAM or MRAM could potentially replace it as it is 4 to 10 times denser which would make it more cost effective.

In addition to the SRAM, some microcontrollers also have internal EEPROM for data storage; and even ones that do not have any (or not enough) are often connected to external serial EEPROM chip (such as the BASIC Stamp) or external serial flash memory chip.

A few recent microcontrollers beginning in 2003 have "self-programmable" flash memory.

Firmware

The earliest microcontrollers used mask ROM to store firmware. Later microcontrollers (such as the early versions of the Freescale 68HC11 and early PIC microcontrollers) had quartz windows that allowed ultraviolet light in to erase the EPROM.

The Microchip PIC16C84, introduced in 1993, was the first microcontroller to use EEPROM to store firmware. In the same year, Atmel introduced the first microcontroller using NOR Flash memory to store firmware.

Cruise Control

Cruise control (sometimes known as **speed control** or **autocruise**, or **tempomat** in some countries) is a system that automatically controls the speed of a motor vehicle. The system is a servomechanism that takes over the throttle of the car to maintain a steady speed as set by the driver.

Icon for cruise control as commonly presented on dashboards

Cruise control mounted on a 2000 Jeep Grand Cherokee steering wheel

Cruise control on Citroën Xsara.

History

Speed control with a centrifugal governor was used in automobiles as early as 1900 in the Wilson-Pilcher and also in the 1910s by Peerless. Peerless advertised that their system would "maintain speed whether up hill or down". The technology was adopted by James Watt and Matthew Boulton in 1788 to control steam engines, but the use of governors dates at least back to the 17th century. On an engine the governor adjusts the throttle position as the speed of the engine changes with different loads, so as to maintain a near constant speed.

Modern cruise control (also known as a speedostat or tempomat) was invented in 1948 by the inventor and mechanical engineer Ralph Teetor. His idea was born out of the frustration of riding in a car driven by his lawyer, who kept speeding up and slowing down as he talked. The first car with Teetor's system was the 1958 Imperial (called "Auto-pilot") using a speed dial on the dashboard. This system calculated ground speed based on driveshaft rotations off the rotating speedometer-cable, and used a bi-directional screw-drive electrimotor to vary throttle position as needed.

A 1955 U.S. Patent for a "Constant Speed Regulator" was filed in 1950 by M-Sgt Frank J. Riley. He installed his invention, which he conceived while driving on the Pennsylvania Turnpike, on his own car in 1948. Despite this patent, the inventor, Riley, and the subsequent patent holders were not able to collect royalties for any of the inventions using cruise control.

In 1965, American Motors (AMC) introduced a low-priced automatispeed control for its large-sized cars with automatitransmissions. The AMC "Cruise-Command" unit was engaged by a push-button once the desired speed was reached and then the throttle position was adjusted by a vacuum control directly from the speedometer cable rather than a separate dial on the dashboard.

Daniel Aaron Wisner invented "Automotive ElectroniCruise Control" in 1968 as an engineer for RCA's Industrial and Automation Systems Division in Plymouth, Michigan. His invention described in two patents filed that year (US 3570622 & US 3511329), with the second modifying his original design by debuting digital memory, was the first electronidevice in controlling a car. Two decades passed before an integrated circuit for his design was developed by Motorola. as the MC14460 Auto Speed Control Processor in CMOS. The advantage of electronispeed control over its mechanical predecessor was that it could be integrated with electroniaccident avoidance and engine management systems.

Following the 1973 oil crisis and rising fuel prices, the device became more popular in the U.S. "Cruise control can save gas by avoiding surges that expel fuel" while driving at steady speeds. In 1974, AMC, GM, and Chrysler priced the option at $60 to $70, while Ford charged $103.

Operation

The driver must bring the vehicle up to speed manually and use a button to set the cruise control to the current speed.

The cruise control takes its speed signal from a rotating driveshaft, speedometer cable, wheel speed sensor from the engine's RPM, or from internal speed pulses produced electronically by the vehicle. Most systems do not allow the use of the cruise control below a certain speed - typically around 25 mph (40 km/h). The vehicle will maintain the desired speed by pulling the throttle cable with a solenoid, a vacuum driven servomechanism, or by using the electronisystems built into the vehicle (fully electronic) if it uses a 'drive-by-wire' system.

All cruise control systems must be capable of being turned off both explicitly and automatically when the driver depresses the brake, and often also the clutch. Cruise control often includes a memory feature to resume the set speed after braking, and a coast feature to reduce the set speed without braking. When the cruise control is engaged, the throttle can still be used to accelerate the car, but once the pedal is released the car will then slow down until it reaches the previously set speed.

On the latest vehicles fitted with electronithrottle control, cruise control can be easily integrated into the vehicle's engine management system. Modern "adaptive" systems include the ability to automatically reduce speed when the distance to a car in front, or the speed limit, decreases. This is an advantage for those driving in unfamiliar areas.

The cruise control systems of some vehicles incorporate a "speed limiter" function, which will not allow the vehicle to accelerate beyond a pre-set maximum; this can usually be overridden by fully depressing the accelerator pedal. (Most systems will prevent the vehicle accelerating beyond the chosen speed, but will not apply the brakes in the event of overspeeding downhill.)

On vehicles with a manual transmission, cruise control is less flexible because the act of depressing the clutch pedal and shifting gears usually disengages the cruise control. The "resume" feature has to be used each time after selecting the new gear and releasing the clutch. Therefore, cruise control is of most benefit at motorway/highway speeds when top gear is used virtually all the time.

Advantages and Disadvantages

Some advantages of cruise control include:

- Its usefulness for long drives (reducing driver fatigue, improving comfort by allowing positioning changes more safely) across highways and sparsely populated roads.

- Some drivers use it to avoid subconsciously violating speed limits. A driver who otherwise tends to subconsciously increase speed over the course of a highway journey may avoid speeding.

However, when used incorrectly cruise control can lead to accidents due to several factors, such as:

- speeding around curves that require slowing down

- rough or loose terrain that could negatively affect the cruise control controls

- rainy or wet weather could lose traction

Adaptive cruise control

Some modern vehicles have adaptive cruise control (ACC) systems, which is a general term meaning improved cruise control. These improvements can be automatibraking or dynamiset-speed type controls.

AutomatiBraking Type: The automatibraking type use either a radar or laser setup to allow the vehicle to keep pace with the car it is following, slow when closing in on the vehicle in front and accelerating again to the preset speed when traffiallows. Some systems also feature forward collision warning systems, which warns the driver if a vehicle in front—given the speed of both vehicles—gets too close (within the preset headway or braking distance).

DynamiSet Speed Type: The dynamiset speed uses the GPS position of speed limit signs, from a database. Some are modifiable by the driver. At least one, Wikispeedia, incorporates crowdsourcing, so driver input is shared, improving the database for all users.

Non-Braking Type: The speed can be adjusted to allow trafficalming. One visual method uses OpenCV

References

- Dyer, S. A.; Harms, B. K. (1993). "Digital Signal Processing". In Yovits, M. C. Advances in Computers. 37. Academic Press. pp. 104–107. doi:10.1016/S0065-2458(08)60403-9. ISBN 9780120121373.

- Liptak, B. G. (2006). Process Control and Optimization. Instrument Engineers' Handbook. 2 (4th ed.). CRC Press. pp. 11–12. ISBN 9780849310812.

- Augarten, Stan (1983). The Most Widely Used Computer on a Chip: The TMS 1000. State of the Art: A Photographic History of the Integrated Circuit. New Haven and New York: Ticknor & Fields. ISBN 0-89919-195-9. Retrieved 2009-12-23.

- David Harris & Sarah Harris (2012). Digital Design and Computer Architecture, Second Edition, p. 515. Morgan Kaufmann. ISBN 0123944244.

- "Oral History Panel on the Development and Promotion of the Intel 8048 Microcontroller" (PDF). Computer

History Museum Oral History, 2008. p. 4. Retrieved 2016-04-04.

- Ingrid Verbauwhede; Patrick Schaumont; Christian Piguet; Bart Kienhuis (2005-12-24). "Architectures and Design techniques for energy efficient embedded DSP and multimedia processing" (PDF). rijndael.ece.vt.edu. Retrieved 2014-06-11.

- "Speak & Spell, the First Use of a Digital Signal Processing IC for Speech Generation, 1978". IEEE Milestones. IEEE. Retrieved 2012-03-02.

- Edwards, Robert (1987). "Optimizing the Zilog Z8 Forth Microcontroller for Rapid Prototyping" (PDF). Martin Marietta: 3. Retrieved 9 December 2012.

Permissions

Index

www.ingramcontent.com/pod-product-compliance
Lightning Source LLC
Chambersburg PA
CBHW061241190326
41458CB00011B/3552